Himalayan Frontiers of India

The Himalayas, a great natural frontier for India, symbolize India's spiritual and national consciousness. The Himalayan region displays a wide diversity of cultural patterns, languages, ethnic identities and religious practices. Along the Himalayas converge the boundaries of South and Central Asian countries, which lend a unique geo-political and geo-strategic importance to this region.

This book provides a comprehensive analysis of historical, geo-political and strategic perspectives on the Himalayan frontiers of India. Drawing on detailed analyses by academics and area specialists, it explains the developments in and across the Himalayas and their implications for India. Topics such as religious extremism, international and cross-border terrorism, insurgency, and drugs and arms trafficking are discussed by experts in their respective fields.

Himalayan Frontiers of India will be of interest to scholars in South and Central Asian studies, International Relations and Security Studies.

Professor K. Warikoo is Director of the Central Asian Studies Programme at Jawaharlal Nehru University, New Delhi. He has authored several books on Kashmir, Central Asia and Afghanistan, and is the founding editor of *Himalayan and Central Asian Studies*.

Routledge Contemporary South Asia Series

1 Pakistan
Social and cultural transformations in a Muslim nation
Mohammad A. Qadeer

2 Labor, Democratization and Development in India and Pakistan
Christopher Candland

3 China–India Relations
Contemporary dynamics
Amardeep Athwal

4 Madrasas in South Asia
Teaching terror?
Jamal Malik

5 Labor, Globalization and the State
Workers, women and migrants confront neoliberalism
Edited by Debdas Banerjee and Michael Goldfield

6 Indian Literature and Popular Cinema
Recasting classics
Edited by Heidi R. M. Pauwels

7 Islamist Militancy in Bangladesh
A complex web
Ali Riaz

8 Regionalism in South Asia
Negotiating cooperation, institutional structures
Kishore C. Dash

9 Federalism, Nationalism and Development
India and the Punjab economy
Pritam Singh

10 Human Development and Social Power
Perspectives from South Asia
Ananya Mukherjee Reed

11 The South Asian Diaspora
Transnational networks and changing identities
Edited by Rajesh Rai and Peter Reeves

12 Pakistan–Japan Relations
Continuity and change in economic relations and security interests
Ahmad Rashid Malik

Himalayan Frontiers of India

Historical, geo-political and strategic perspectives

Edited by
K. Warikoo

Routledge
Taylor & Francis Group
LONDON AND NEW YORK

First published 2009 by Routledge
2 Park Square, Milton Park, Abingdon, Oxon OX14 4RN

Simultaneously published in the USA and Canada
by Routledge
711 Third Avenue, New York, NY 10017

Routledge is an imprint of the Taylor & Francis Group, an informa business

First issued in paperback 2012

Typeset in Times New Roman by
RefineCatch Limited, Bungay, Suffolk

British Library Cataloguing in Publication Data
A catalogue record for this book is available from the British Library

Library of Congress Cataloging-in-Publication Data
Himalayan frontiers of India : historical, geo-political, and strategic perspectives / [edited by] K. Warikoo.
p. cm.—(Routledge contemporary South Asia series ; 13)
Includes bibliographical references and index.
1. Geopolitics—India. 2. Geopolitics—Himalaya Mountains Region. 3. India—Foreign relations. I. Warikoo, K. (Kulbhushan), 1951–
DS485.H6H54995 2009
327.54—dc22 2008021719

ISBN13: 978–0–415–46839–8 (hbk)
ISBN13: 978–0–203–88732–5 (ebk)
ISBN13: 978–0–415–53349–2 (pbk)

Contents

Contributors

Prof. K. Warikoo is Director of the Central Asian Studies Programme, School of International Studies, Jawaharlal Nehru University, New Delhi.

Maj. Gen. (Rtd.) Afsir Karim is the editor of *AAKROSH*, quarterly journal on terrorism and internal conflicts.

Prof. Vijay Kapur is Professor in the Faculty of Management Studies, University of Delhi, Delhi.

Sat Paul Sahni, a veteran journalist from Jammu and Kashmir and former Director General of Information and Public Relations, J and K Government, is author of several books on Kashmir.

P. N. Jalali, a veteran journalist from Jammu and Kashmir, was Chief of PTI, Srinagar for more than 20 years.

B. Raman is former Special Secretary, Cabinet Secretariat, New Delhi.

Prof. B. C. Upreti is Director of the Centre of South Asian Studies, University of Rajasthan, Jaipur.

Prof. P. Stobdan is former Director of the Centre for Regional and Strategic Studies, University of Jammu, Jammu.

Dr Rajesh Kharat is Associate Professor in the Department of Political Science, University of Mumbai, Mumbai.

M. M. Khajooria is former Director General of Police, Jammu and Kashmir.

B. B. Nandy is former Special Secretary, Cabinet Secretariat, New Delhi.

Air Marshal (Rtd) Vinod Patney is former Chief of the Air Staff.

Preface

The Himalayan range is the embodiment of divinity, of nature in its splendour and of culture in the deepest sense of the word. It has been inextricably interwoven with the life and culture of India since time immemorial. It has been the repository of rich biodiversity, the source of main river systems and glaciers and the symbol of India's spiritual and national consciousness. It is the geographical feature that dominates India most and which has acted as a great natural frontier. Though geographically speaking the Himalayan Range is embraced at its western and eastern extremities by the Indus and Brahmaputra respectively, we cannot isolate the Hindu Kush, Karakoram and Pamirs regions, which are continuous and interlocked with the great Himalayan mountain system. Stretching over 2,500 km from Kashmir in the west to Arunachal Pradesh in the east, it has provided India with a natural and most formidable line of defence. However, its imposing geographical features did not prevent the region from being a complex of cultural interaction, migration, overland trade and communication. The Himalayan region has been the cradle from where ancient Indian culture, including Mahayana Buddhism, spread to different countries in Central, South East and East Asia. Such cross-cultural contacts were not confined only to the religious philosophy of Mahayana Buddhism, but also included art, architecture, literature etc. The movement produced a harmonious blend of cultures, arts, science and literature. After the Chinese occupation of Tibet, Indian Himalaya became the last refuge of Buddhism. That explains the rationale behind the setting up of specialized Buddhist Studies Institutes in Ladakh, Gangtok and Arunachal Pradesh after the late 1950s.

The importance of the Himalayas as the natural frontier of India in the north is immersed in Indian ethos and psyche.

अस्त्यूत्तरस्याँ दिशि देवात्मा हिमालयो नाम नगाधिराजः ।

पूर्वापरौ तोयनिधी वगाह्य स्थितः प्रथिव्या इव मानदण्डः ।।

[In the north (of our country) stands the Lord of Mountains and the very embodiment of divinity – the Himalayas, like a measuring rod of the earth spanning the eastern and western oceans.]

This is how Kalidasa in his *Kumara Sambhava* described the Himalayas as *devatma* – a divine personality, and as the measuring rod spanning the Eastern and Western Oceans, thereby pinpointing the northern frontiers of India. To the majority of Indians, the Himalayas are mythical mountains referred to by the Vedas, Puranas and other scriptures. The Himalayas are part of our history, tradition and cultural heritage. Most of our sacred shrines and places of pilgrimage are situated in the Himalayan heights. So much so, there is no fulfilment of life to an Indian without some sort of Himalayan experience.

The very fact that the boundaries of Tajikistan, Afghanistan, China, Pakistan, India and Myanmar converge along the Himalayas lends a unique geo-strategic importance to this region. Its potential for instability and conflict is furthered by the ethnic-religious jigsaw prevailing in the Himalayas and trans-Himalayas where people of Buddhist, Hindu and Islamic faiths are concentrated in various areas and are vulnerable to extraneous influences. Major international land frontier disputes pertain to this area. Whereas India and Pakistan have been locked in a conflict over Kashmir since 1947, the Sino-Indian border dispute remains to be settled. Any cross-border fraternization of people of Xinjiang and Tibet in China, Uzbekistan, Tajikistan, Afghanistan, Pakistan, the Indian Himalayas from Kashmir up to North East India, Nepal, Bhutan, Bangladesh and Myanmar, on ethnic and religious lines is a potential source of conflict in the region and a threat to the security of concerned states. With the disintegration of the erstwhile USSR and the emergence of newly independent Central Asian states – all having a predominantly Muslim population, a new geo-political situation has arisen across the north-western Himalayas. Due to its geo-strategic proximity to South and West Asia, Central Asia has emerged as a distinct geo-political entity stimulating global attention and interest.

The rise of the Taliban to power in Kabul in September 1996, which turned Afghanistan into the centre of Islamist extremism, global terrorism, and drugs and arms trafficking brought Central Asia to the focus of global attention. The establishment of an extremist Islamist order in Afghanistan and the active involvement of Islamist *Mujahideen* in cross-border terrorism and *jihad* (Holy war), whether in the Indian state of Jammu and Kashmir, Tajikistan or some other CIS (Commonwealth of Independent States) countries, adversely affected regional security, peace and stability in the Himalayan and adjoining Central Asian region. Though the Taliban and Osama bin Laden and his network were actively engaged in encouraging Islamist extremism and terrorism in South, Central and South East Asia and also in the West, it was only after the dreadful terrorist strikes on the World Trade Center and Pentagon on 9/11, that the United States and its Western allies realized the severity of the challenge they posed. Even after more than seven years of global campaigning against terror, the Taliban and Al Qaeda cadres have not been vanquished. In fact, the past two years have witnessed their resurgence, thereby posing a great challenge to sustainable security and peace in Afghanistan and the adjoining regions. As such, the entire region has become susceptible to forces of religious extremism, cross-border terrorism, insurgency, and drugs and arms trafficking.

The western Himalayas have served as the gateway to India for numerous invasions and influences from Central and West Asia. Even in contemporary times, India has had to experience successive military aggression from Pakistan in 1947, 1965, 1971, 1999 (in Kargil) and now in the form of a proxy war in Kashmir that has continued for more than 18 years, besides the Chinese military offensive in 1962 from across the Himalayas.

Taking the case of Ladakh, it enjoys a unique geo-strategic location, being bounded by Xinjiang in the north, Tibet in the east, Kashmir and Baltistan in the west, and Lahoul, Spiti, Kulu, Bushahr and Chamba in the south. Ladakh has played an important role in the history and culture of this frontier region. Enjoying a central position in the network of overland caravan routes that were linked to the Silk Route it acted as an important gateway in the Indo-Central Asian exchange of men, materials and ideas through the ages. However, during the nineteenth and early twentieth centuries, the extent and pattern of these exchanges was influenced on the one hand by the state of diplomatic relations between the three empires – Britain, Russia and China – and on the other by the level of influence exercised by the Dogra rulers of Jammu and Kashmir. The 1830s witnessed the beginning of the 'Great Game' played by Britain and Russia in Central Asia. In 1834, Ladakh came under Dogra control, and by 1840 their forces, led by General Zorawar Singh, had established their authority throughout Ladakh and Baltistan. In 1846, it became part of the newly founded state of Jammu and Kashmir under the Dogra ruler Maharaja Gulab Singh and his successors. And since 1947, Ladakh, including Kargil, has been a province of the Jammu and Kashmir state, with its borders abutting Xinjiang and the Tibetan regions of China to its north and east, and Baltistan in Pakistan-occupied Kashmir (PoK) to its west.

Since a sizeable portion of Ladakh territory (Baltistan, Raskam, Aksai Chin) had been under the occupation of Pakistan and China since 1947 this remains an important issue of study and research by Indian scholars. It may be pointed out that the old established frontiers on the Tibet–Ladakh border were reaffirmed by the peace treaty signed between the Dogras and Tibetans in September 1842. The Tibetans also accepted the Dogras as the legitimate authority in Ladakh, and trade in shawl-wool and tea was continued in accordance with old customs via Ladakh. Similarly, Baltistan (Skardo) which was later occupied by Pak forces in 1947–48, has ever since been part of the Pak-occupied Kashmir/Northern Areas. However, Lahoul and Spiti, which were part of Raja Gulab Singh's territory of Ladakh, were taken away by the British in 1846–47, and merged with the British possession of Kangra in the western Himalayas. Ladakh's distinct geo-cultural identity was consolidated by the Dogras who maintained a separate *Wazarat* for Ladakh and Baltistan.

It maintained social, cultural and economic linkages with western Tibet and also with the neighbouring principalities of Lahoul, Spiti, Kulu and Bushahr. The history of Ladakh's control over Gartok and Rudok in western Tibet is a matter of importance. Minsar, an enclave of the Dogras within western Tibet, used to pay revenue to the Dogras until the early twentieth century. The British used Ladakh and adjoining areas in Gilgit, Skardo, Hunza and Chitral as 'frontier listening posts' to monitor developments in Central Asia and Xinjiang throughout the Dogra period.

Issues like Pakistan's aggression in Ladakh and Baltistan in 1947–48, the heroic resistance by Ladakhis, repulsion of the Pak invasion, and also the circumstances leading to the non-recovery of Baltistan/Skardo from Pak control, became even more relevant during the Pakistani intrusion in Kargil in 1999, which further underscored the need for effective security management of the Indian Himalayan frontiers. The Kargil crisis unleashed latent nationalism and fierce patriotism throughout India, cutting across regional, religious and caste barriers. The people of India demonstrated exemplary courage, a deep-rooted sense of national unity and a commitment to the territorial integrity of India, and resolve and willingness to sacrifice their lives to defend their Himalayan borders. It also brought into focus the vulnerability of Indian positions in Kashmir along the Srinagar–Dras–Kargil–Leh strategic highway, due to Pakistan controlling the heights overlooking this highway. It only shows how unprepared Indian military and foreign office authorities were at the time of delineation, first of the ceasefire line (CFL) in 1949, and later the Line of Control (LoC) in 1971–72. Successive political leaders have allowed the battle to be lost at the negotiating table even if it was won at the battle front at high human costs.

In the early twentieth century, Lord Curzon had visualized the Himalayan regions of Ladakh, Sikkim, Bhutan and North East frontier as an 'inner defence line for India protected by a Tibetan buffer region'. Later, China viewed the Himalayas as its outer line of defence, necessary for the protection of its interests in Tibet. Some analysts felt that Communist China considered the high plateau of Tibet to be the palm of the hand with Ladakh, Nepal, Bhutan, Sikkim and the North East Frontier Agency as the five fingers. China has the palm under its control and now it seeks influence over the strategic five fingers.

The importance of the Himalayas as the protective barrier for India is enhanced by the vast expanse of the Tibetan plateau, whose elevation is above that of ordinary mountain ranges. In the words of K. M. Pannikar, 'the plateau of Tibet has Kuenlun mountains as its boundary in the north, the Karakoram in the west and south, and an equally mountain-bound boundary in the east. The mountainous area to the north of India has to be considered strategically as one great quadrilateral, the middle of which is an elevated plateau of 15,000 feet (c 460 metres) above sea level, and the southern ramparts, an invulnerable range of an average height of 20,000 ft (c 610 m). The area enclosed is 500,000 square miles (c 93,050 km^2), frightening and formidable in its geographical features, an arid waste, wind-swept and waterless where trees do not grow'. So in spite of the new and sophisticated advancements in weapon and communication technologies, the traditional importance of the Himalayas as the land frontier and natural barrier has not been diminished. As Pannikar emphasized, 'the essential point about the Himalayas is not their width of 150 miles (c 240 km), but the plateau behind it, which in itself is an elevation of about 15,000 ft (c 460 m) and is guarded on all four sides by high mountains. In fact, the vast barrier upland behind the Himalayas provides the most magnificent defence in depth imaginable.'

The Tibet question has remained the focus of international and national attention with India now hosting Tibetan refugees for over four decades. Whereas India has been generous to the Tibetan refugees, providing land for settlement and avenues for their education and sustenance, it has not used their presence or that of the Tibetan Government in Exile (TGE) at Dharamshala, as a bargaining chip in its dealings with China. This is despite the fact that the pro-Tibetan and pro-Western lobbies in India have been active in their sustained campaign impressing upon the Government of India the need to lend political support to the Tibetans' demand for independence and recognition of a *TGE*. For the past few years, there has been noticeably greater activity in the involvement of some Buddhists of the Indian Himalayas, particularly in Ladakh, Himachal Pradesh, Sikkim and Arunachal Pradesh, to press for this demand. Several forums, such as the Himalayan Committee for Action on Tibet, the Himalayan Buddhist Cultural Association, the Tibet Sangharsh Samiti etc., have been formed, with their branches operating in all the states of the Indian Himalayas. These forums have been pressing upon the Government of India to support the Dalai Lama's proposals on Tibet and also to recognize the TGE headed by him. Besides that, demands have been made to include the Bhoti language in Schedule VIII of the Indian constitution, and also to provide for the preservation and promotion of Tibetan and Himalayan art and culture.

The above-mentioned facts need to be viewed in the light of sustained efforts by Tibetan scholars working in Dharamshala or in the west towards preparing a unified system of the Tibetan language so that the same script and dialect etc., is applicable to all Tibetan speaking peoples whether in India, Tibet or elsewhere. Some Western scholars have advanced the theory of Shambala World, advocating unification of non-Chinese, non-Russian and non-Indian races in the Central Asian, Himalayan and trans-Himalayan region. This raises the question of the Tibetanization of society, culture and politics of the Indian Himalayas particularly in Ladakh, Himachal Pradesh, Sikkim, Arunachal Pradesh etc. It has been noticed that Tibetan settlers in these parts do not use local dialects and seek to exercise their cultural superiority over the local Buddhist inhabitants, as they harbour an exalted view of Tibetan culture. Due to divergent modes of economic activity being followed by the Tibetan settlers and the indigenous Buddhists in the Indian

Himalayas, the former being engaged in marketing and business activities and the latter being involved in a primarily agrarian economy, there have been social conflicts between these two culturally similar groups, with the locals viewing the Tibetan settlers as exploiters.

As a Ladakhi Buddhist scholar points out:

> 'the intense Tibetanization of the Himalayan region over the last two to three decades seems to have served China's interests well. The Dalai Lama has cleverly carved out space for himself through cultural and religious activities along the Himalayan belt from Arunachal Pradesh to Ladakh. To add to the list of dozens of Tibetan spiritual leaders having a foothold in the Himalayas, China is exporting more influential lamas into India, the most prominent of them being the Karmapa Lama. The fact that the sixth Dalai Lama was born in Tawang, is being added to China's new articulation of its claim over Arunachal Pradesh. There is no doubt that China's "Western Development Campaign" would enhance the scope for its influence across the Himalayas. The picture is getting increasingly confused along the Sino-Indian frontiers, and it could be that a solution to the Tibet problem would be found only at India's expense.'

[H]ence the need to settle the Sino-Indian border dispute in a manner that takes into account long-term strategic and defence requirements of India. Whereas the political frontiers should be defined and delineated, Indian diplomacy and political leadership should ensure that such frontiers are not breached by any overt or covert operations by the enemy. India and China, being the largest populated countries comprising nearly 40 per cent of the total humanity, are poised to achieve greater heights in their economic, industrial, technological and military prowess in the twenty-first century. Rich in human and material resources and having a wide diversity of peoples and cultures, both countries share a long history of historico-cultural interface. In the post-Cold War era, both India and China can play a constructive role in maintaining peace and security in South and Central Asia, besides enhancing the prospects of bilateral trade, commerce and human development in these two countries.

Both India and China share similar views on major issues, particularly economic development, pursuit of economic, social and cultural rights, threats posed by drugs and arms trafficking, trans-border terrorism, religious extremism and ethnic-religious separatism of the territorial integrity of nation states. Yet, there remain substantial differences on the issues of the Sino-Indian boundary and China's military and nuclear assistance to Pakistan, which continue to be a matter of concern in India. Though China claims that the boundary question is a legacy of the British rule in India, the fact remains that China has resolved its similar long-standing boundary disputes with Mongolia, Russia, Kazakhstan and Kyrgyzstan. In these cases too, China had based its claims on the same premise that these border disputes were a legacy of the imperialist Tsars. Therefore, there seems to be no plausible reason as to why the Sino-Indian border issue cannot be resolved after mutually satisfactory negotiations.

The former Indian Prime Minister, Atal Behari Vajpayee, in his address to the Combined Commanders' Conference on 1 November 2003, was quite forthright in underlining the need for a pragmatic approach by India to resolve the Sino-Indian boundary question finally. To quote Vajpayee:

> 'our border with China has remained largely peaceful for the past few decades. During my recent visit to China, we agreed to raise our bilateral and economic cooperation to a qualitatively higher level. The decision to appoint special representatives to discuss the boundary question from a political perspective was a particularly significant measure. A final resolution of the boundary question would release considerable military energies and finances for more purposeful activities. It is, therefore, a strategic objective. To achieve it, we should be willing to take pragmatic decisions.'

Of late, India has woken up to the need for removing the existing bottlenecks in road, rail and communication linkages between the mainland of India and its Himalayan frontier outposts and even beyond in order to turn the entire frontier belt into a bridge of friendship and cooperation.

It is against this background that this book deals with both the historical, geo-political and strategic perspectives on the 'Himalayan Frontiers of India' so that one gets a holistic and updated view of the developments in and across the Himalayas and their implications for India. The book is based on the papers presented by eminent academics, area specialists and experts in border management, including military officers, former heads of police, the border security force and the Indian Air Force, at a seminar organized by the Himalayan Research and Cultural Foundation more than two years ago.

Chapters 1–3 present historico-political perspectives This author's first paper provides a rounded view of India's trans-Himalayan trade and cultural contacts with Central Asia through Kashmir and Ladakh (1846–1947) and his second gives fresh insights into the 'Great Game' on Kashmir frontiers during that period. P. N. Jalali, an octogenarian journalist and political activist from Kashmir, sheds important light on the Gilgit dimension of the Kashmir frontier, providing something of a contemporaneous account.

The book then continues by dealing with strategic perspectives. Prof. Vijay Kapur, while analysing the new strategic, political and economic challenges, stresses the need for India to devise a well-integrated policy paradigm for its Himalayan frontier, providing a strategic vision as well as policy imperatives. Major General (Rtd) Afsir Karim brings into focus the strategic dimensions of the trans-Himalayan frontier in the Gilgit-Baltistan region. Sat Paul Sahni, another octogenarian journalist and former Director General of

Information in the Jammu and Kashmir Government, who bases his paper on personal experiences as a war correspondent in Jammu and Kashmir since 1941, traces the evolution of the CFL and LoC into a border. The ground situation in Gilgit-Baltistan, also called the 'Northern Areas' of Pakistan, is analysed by B. Raman. The problem of cross-border terrorism in the Jammu and Kashmir state of India and its *jihadi* context has been well presented by M. M. Khajooria, who, being the former Director General of Police in Jammu and Kashmir, was witness to this phenomenon during its peak period. Professor P. Stobdan, a Ladakhi Buddhist himself, presents a fresh outlook on the linkage between the issues of Tibet, Tibetan refugees in India and the security of India's Himalayan frontiers. The inter-linkages, issues and problems relating to India–Nepal open borders have been dealt with by B. C. Upreti. The entire gamut of Indo-Bhutan relations and the imperatives for the security of India's Himalayan frontiers have been covered by Rajesh Kharat. B. B. Nandy, former Director General of the Border Security Force of India, who has had long experience in dealing with the problem of India's open and porous borders with Bangladesh and Myanmar, has provided his insights into the issues of illegal migration, smuggling of arms and drugs and security in the north-eastern frontiers of India. The book concludes with a chapter by Vinod Patney, former Chief of the Air Staff, who has evaluated the role of science and technology, modern air surveillance and remote sensing as a means to ensure the security of India's Himalayan frontiers.

Acknowledgments

I acknowledge with thanks all the contributors to this book, for sharing their expertise, knowledge and experience through their papers. I am particularly grateful to the Himalayan Research and Cultural Foundation for allowing me to produce this book, which is based on the papers presented at a seminar organized by the foundation, and published by Routledge both in print and electronic editions. Last but not least, I would like to thank Dr. Mahesh for his assistance and Mumtaz Alam for typing the manuscript.

K. Warikoo

1 India's gateway to Central Asia

Trans-Himalayan trade and cultural movements through Kashmir and Ladakh, 1846–1947

K. Warikoo

Introduction

Its contiguity with Central Asia, Xinjiang and Tibet and its proximity to the Silk Route turned Kashmir and Ladakh into a major gateway in the Indo-Central Asian exchange of men and materials through the ages. Enjoying a central position in the network of caravan trails, Leh – the capital town of Ladakh – was an important transit emporium where Indian traders met their counterparts from Tibet, Central Asia and Afghanistan. Undaunted by numerous physical obstacles and displaying a unique adventurous spirit, these traders roamed across the lofty Himalayan and Karakoram mountains and the barren deserts. Usually the traders would exchange or sell their goods at Leh thus making it the terminal point of the multilateral trade carried between India, Tibet and Central Asia. This paper seeks to analyse the extent and pattern of trans-Himalayan trade carried through Kashmir and Ladakh with adjoining territories of Chinese Turkestan (Xinjiang) and Tibet, and also to assess the impact of this trade on the society and culture of Ladakh and Kashmir.

Trade routes

Numerous caravan routes that converged near Leh acted as the channels of communication between India and Central Asia. From the Indian side the bulk of the trade passed through the Srinagar–Leh route, though traders from Skardu, Kishtwar, Kullu, Lahoul, Spiti, Nurpur and Bushahr used to carry on their business with Ladakh directly through the Khapalu–Chorbat–Nobra, Kishtwar–Zanskar and Kullu–Rohtang passes, Lahoul–Key Long–Bara Lacha pass, Rupshu–Longa Lacha pass and Thung Lung La and Leh routes respectively. To the east of Ladakh, a caravan trail passed through Gartok towards Lhasa, the capital of Tibet. In the north, Ladakh stood connected with the Silk Route at Yarkand by a caravan route running across the Karakoram pass. From Leh there existed three paths leading to Yarkand, the nearest big trading centre of Central Asia. Those traders and passersby who opted to travel to Yarkand in winter would cross Digar La and follow the narrow and winding valleys of the Shyok river. This river, which was frozen during winter, was to be crossed and re-crossed several times. After negotiating the Karakoram pass, traders would start their descent towards Yarkand passing through Kugiar and Karghalik. But it was usual for the Central Asian and Indian traders to bring their caravans to Ladakh in summer and start their return journey homewards in the autumn. The summer route from Leh to Yarkand passed through Khardung La, Nobra valley, Saser La, Karakoram pass and the Suget pass. Yet another route ran across the Changla pass, the Chang Chenmo valley and Lingzithang plains to join the Leh–Yarkand summer route at Shahidulla. The Chang Chenmo valley route to Eastern Turkestan was improved considerably after the conclusion of a treaty between Kashmir Durbar and the British Indian Government in 1870. Under the treaty, this route was also declared a free highway. Efforts were made to establish this track as the main trade route between Leh and Yarkand, as it was easier than those by the Shyok or the Saser pass. But being relatively longer in distance and fuel, grass and water not being so abundantly available on this route, it did not become so popular with the traders, excepting those who used camels to transport their goods.

Notwithstanding its physical difficulties, the Srinagar–Leh–Yarkand route was the most important and longest established thoroughfare between India and Central Asia. Even though the British authorities took numerous steps to improve the Kullu–Leh route with a view to encouraging direct trading between British India and Central Asia, traders continued to use the Srinagar–Leh route as it was 'the easiest and best supplied as to grass, provisions, etc., and thoroughly open for two or three months longer than the Kullu route'.[1] It was through this route that the Kashmir shawl industry received its supplies of pashm wool that was imported into Ladakh from Tibet and Yarkand. Soon after his conquest of Ladakh, the Dogra General, Wazir Zorawar Singh, improved this route to make it an easy passage for mounted travellers.[2] The road was kept in excellent repair throughout the period of Dogra rule in Kashmir. As a sequel to the treaty concluded between Maharaja Ranbir Singh and T. D. Forsyth in April 1870, the Kashmir Government allocated a yearly sum of 1,500 rupees for maintenance of roads and sarais along the Indo-Central Asian trade route which came to be known as the Treaty Road. In the late nineteenth century an extra annual grant of 5,000 rupees was made available by the Kashmir Durbar for this purpose.[3] There were two big caravan-sarais each at Leh and Srinagar for accommodating the traders and Central Asian pilgrims. In addition, godowns, stables, dak bungalows and inns were established at various stages of the route where traders received shelter and supplies of grain and forage for the ponies at subsidized rates. Similarly, the villagers living in and around a particular stage of the trade route in Ladakh were responsible for supplying pony and coolie transport to traders and travellers at reasonable rates. It became known as the *Res* system under which 15 to 20 ponies would always be available at each stage for use by the traders, public servants and travellers enjoying official patronage.[4] Such was the safety of this route that in the event of unfavourable weather or death of ponies, traders would march on to the next safe stage leaving behind their goods, which were fetched after the climate became favourable or substitute transport became available. As a result, the Srinagar–Leh–Yarkand route became the most important thoroughfare between India and Central Asia during the Dogra rule in Kashmir.

The extent and pattern of trade

Ladakh's place in the Indo-Central Asian trade was only that of a transit emporium, as it did not produce or consume much to make any indigenous trade of importance. Traders from Tibet, Central Asia and India gathered at Leh to exchange their merchandise. This commercial intercourse not only sustained the poor and backward economies of the semi-closed systems in these remote and high-altitude areas, but also provided the essential raw materials for the flourishing shawl and carpet industry of Kashmir. It also lent strong support to the local trade passing between Leh and Skardu (Baltistan) which formed a *Tehsil* of the Ladakh Wazarat of the state of Jammu and Kashmir during the Dogra rule.[5]

Local trade with Baltistan

The domestic trade between Leh and Skardu was similar to what was carried between Leh and adjoining areas like Nobra, Zanskar, Puring and Chang Thang. It passed through two channels – Leh-Indus Valley–Skardu and Leh–Nobra–Chorbat–Khapalu. A substantial number of Baltis would visit Leh throughout the year. Balti peasants and petty traders brought for sale to Ladakh the produce of their farms and households such as apricots, apricot oil, butter, grapes, almonds, barley, teacups, stone vessels for cooking, woollen cloth and coarse shawls. They took back in exchange Indian cotton cloth and other piece goods, tea, gold and silver thread, Yarkandi leatherware, sheep wool and also some Yarkandi and Tibetan wool which was used in the manufacture of Balti shawls. From Leh these apricots and coarse shawls were transmitted to Lhasa and Yarkand through the medium of regular trade channels as these were in great demand there. Between 400 to 500 maunds of dried apricots from Baltistan were exported to Tibet each year until the early 1870s and later on its quantity increased to about 1,500 maunds per year.[6] Some adventurous Baltis went to Jammu and the plains of Punjab and even as far as Yarkand (in Chinese Turkestan) in search of livelihood and stayed there for a few years to work as manual labour or do small trade. On their return home these Baltis used to invest their savings in purchasing some merchandise which they carried to and sold in Baltistan. Unemployed Baltis would earn their livelihood by working as porters and pony drivers for traders. The trans-Himalayan trade that passed through Ladakh further integrated its economy with that of Baltistan which was already under its administrative jurisdiction, so much so that many Kashmiris set up shops in the Skardu bazaar and engaged themselves in weaving shawls for which pashm wool was brought to Ladakh.[7]

Trade with Tibet

Tibet's trade with Ladakh and Kashmir was regulated by the Treaty of Tingmosgang, concluded in 1684, under which Ladakh got the monopoly over shawl-wool produced in Tibet, and the Tibetans acquired the exclusive right to the brick-tea trade with Ladakh.[8] The treaty also provided for the despatch of periodic missions by Ladakh to Lhasa carrying presents for the Dalai Lama.[9] Since the bearers of this religious mission were allowed to carry merchandise, it soon acquired a commercial character. The shawl-wool that was imported from Tibet and Xinjiang into Ladakh was exclusively exported to Kashmir through the agency of Kashmiri merchants. Brick-tea imported from Tibet was almost entirely consumed within Ladakh and also forwarded to Kashmir. Such trade ties between Ladakh and Tibet were reinforced by the Treaty of 1842 concluded between the Dogra and Tibetan officials. Under this treaty semi-official trade missions were exchanged at intervals between Leh and Lhasa.[10] The triennial mission which left Leh for Lhasa carrying presents and merchandise was known as the *Lapchak* mission. The annual Tibetan caravan coming to Leh with brick-tea and other goods was called *Chaba*. Whereas the Ladakhi traders were entitled to free transport and accommodation during their travel and stay in Tibet, traders from Tibet enjoyed similar facilities during their sojourn in Ladakh. After the final annexation of Ladakh by the Dogras in 1842, the *Lapchak* mission lost its religious character. It was now being managed by the professional traders – mainly Kashmiri Muslims settled in Leh – who earned large profits in the transactions. When Sven Hedin visited the well-known Muslim merchant Haji Nazer Shah at Leh during a stopover in his journey to Tibet in 1906, he was astonished to see in his commercial house 'chests full of silver and gold dust, turquoise and coral, materials and goods' to be sold in Tibet.[11] He soon found that the source of this wealth was their monopoly over the *Lapchak* mission which fetched them an annual profit of about 25,000 rupees.[12] To this Muslim family of about one hundred members headed by Nazer Shah, was entrusted 'the duty of carrying out the *Lapchak* mission' by the Maharaja of Kashmir and at the time of Sven Hedin's visit in 1906 they had retained this confidential post 'for some fifty years'.[13] The close involvement of Kashmiri Muslim settlers in Ladakh's trade with Tibet can be gauged from the fact that even as late as 1959, about 129 such families having some 600 members were residing in the Lhasa-Shigatse area of Tibet.[14]

The Ladakhi imports from Tibet comprised fine pashm wool, coarse wool, brick-tea, salt, silver, gold, turquoise, teacups, paper and musk.[15] Strachey, who visited Ladakh in 1847, estimated the value of such imports as one and a quarter lakh rupees.[16] Ladakh's exports to Tibet included silver ingots (*yambus*), gold, China silk and coarse cotton goods received from Eastern Turkestan, glassware, coral, silver coins, cotton cloths, chintzes, brocades, goatskins and furs imported from British India, saffron and rice from Kashmir, and apricots and barley from Ladakh and Baltistan.[17] Apricots, Chinese silver, Turkestan-made cotton goods and silk cloth, Kashmir saffron and grains constituted the bulk of Ladakhi exports to Tibet. Much of this trade was generated by the heavy demand for pashm wool in Kashmir which had developed into the main centre of shawl production. The assured supplies of pashm wool from Tibet not only sustained Kashmir's shawl-industry, but also provided new avenues of employment to skilled and unskilled workers in Ladakh and Kashmir. Tibet was dependent on Ladakh and Kashmir for supplies of essential goods like food grains, saffron, shawls, coarse cotton cloth, leather and apricots. That the cotton and silk fabrics and carpets produced in Kashgar and Khotan were also transmitted to Tibet through Ladakh, shows that there existed no direct communication between western Tibet and Eastern Turkestan during this period. The annual turnover of Ladakh's trade with Tibet during the Dogra rule amounted to several lakh (hundred thousand) rupees. This trade suffered heavily after the incorporation of Tibet in the People's Republic of China in 1950. It continued to operate however, though under strain, until it came to a standstill in 1959 following disturbances in Tibet.

Indo-Central Asian trade in Ladakh and Kashmir

The bulk of Indian trade with the Central Asian towns of Yarkand, Kashgar and Khotan was carried through Kashmir and Ladakh. Central Asian exports to Ladakh and Kashmir comprised gold and silver, hemp drug, shawl-wool, carpets and felts, tea, Chinese teacups, leather ware, coarse cotton cloths, raw silk and ponies. Out of these items bullion, cannabis (*charas*) and shawl-wool constituted the major imports. The Yarkandi and Andijani traders used to bring these goods to Ladakh where they exchanged the same with their Indian counterparts. Occasionally, the Central Asian traders would move forward to Kashmir and Punjab in the hope of realizing better profits. Similarly, Indian traders would also proceed beyond Ladakh towards Yarkand to make direct purchases at relatively lower prices. Indian traders brought to Leh, Indian and British made cotton cloths, brocades, Kashmir shawls, indigo, spices, dyed goat skins, opium, preserved fruits, coral, indigenous medicines, sugar and books. Whereas most of the Indian imports were exported to Yarkand, part of it also went to Lhasa.

In the late 1840s, merchandise valued at about seven and a half lakh rupees exchanged hands in Ladakh each year.[18] Out of this figure, the imports from Yarkand and India via Kishtwar, Nurpur, Kulu and Bushahr represented the amount of one and three-quarter and four and a half lakh rupees respectively.[19] Another contemporary estimate of trade passing through Ladakh has been provided by Cunningham, who visited Leh twice during the years 1846–47. According to him annual Indian imports into Ladakh averaged 2.2 lakh rupees, whereas the Central Asian exports to India via Ladakh were valued at 2.38 lakh rupees.[20] This meant an annual trade turnover of little more than four and a half lakh rupees. Though Cunningham's estimate is much less than that of Strachey, we get a rough assessment of the volume of annual trade passing through Ladakh in the 1840s which can be safely put between six and seven lakh rupees.

The extent and pattern of overland Indo-Central Asian trade carried through Ladakh fluctuated from time to time due to the changing political situation in Central Asia. In the early nineteenth century, when the relations between Omar Khan, the ruler of Kokand, and the Chinese authorities in Eastern Turkestan had become strained, all the Andijani traders left Yarkand for their country and abandoned their business trips to Ladakh.[21] This resulted in a considerable decrease in Yarkandi exports to Ladakh.[22] Such disruptions in trade did occur even afterwards when Eastern Turkestan was rife with turmoil caused by the Khoja uprisings against the Chinese rulers.

Until the late 1860s, Yarkandi exports to Ladakh were mainly in the form of Bukharan and Kokandi gold coins, Khotanese gold dust and jade, Chinese silver ingots, tea, hemp drug, China-silk, Russian leather etc. Chinese silver ingots (*yambus*) were abundantly available in Ladakh at the rate of about 170 rupees each. These were largely exported to Kashmir where they were melted and manufactured into silverware. According to Hugel, an Austrian traveller who visited Kashmir in the 1830s, these silver ingots bearing Chinese stamp markings were made into coins in Kashmir.[23] It was during this period that large quantities of opium produced in Kishtwar and Bushahr were exported to Eastern Turkestan where it was consumed by the Chinese traders, civil and military officials. Despite the Chinese prohibition imposed in 1839 on the import of opium, this drug was exported to Eastern Turkestan by Indian traders to the extent of 210 maunds per year with the connivance of Chinese customs officials.[24] During Yakub Beg's rule in Kashgharia (1867–77), little or no opium was exported there. But soon afterwards, the value of Indian opium exported to Chinese Turkestan during the years 1878–80 reached the figure of 86,000 rupees because the Chinese consumers had reappeared in Kashgharia.[25]

The volume of trade between India and Kashgharia registered a steady increase after the Chinese were driven out by Yakub Beg in 1867. This was because of the stoppage of all Chinese imports into eastern Turkestan and also due to the development of harmonious relations between Yakub Beg and the British Indian authorities. The abolition of all transit duties on merchandise passing to and from Central Asia through Kashmir and Ladakh by Maharaja Ranbir Singh also encouraged this trade. During the rule of Yakub Beg in Kashgharia export of Indian tea to that quarter increased considerably. Even the Chinese tea imported into Ladakh via Lhasa and

Punjab was exported to Eastern Turkestan to meet the shortages caused there by the cessation of its trade with China. British manufactured cotton and silk cloths now began to be a substitute for Indian cotton prints and silks. However, charas, pashm wool and ponies continued to be imported into Ladakh from Eastern Turkestan and as stated above Indian exports to Eastern Turkestan shot up to an unprecedented figure of more than eight lakh rupees during the year 1874.[26] This included arms and ammunition worth 1.2 lakh rupees imported by Yakub Beg for his army.[27] Such an unusual increase in Indian exports was largely due to the fact that the British merchant Russell took a large quantity of English cotton goods and silks to Eastern Turkestan for sale via Ladakh.[28] However, Ladakh's import and export trade with Eastern Turkestan decreased by more than four and a half lakh rupees in 1877 when unsettled political conditions prevailed there on account of a war between Yakub Beg and the Chinese forces.[29]

The death of Yakub Beg in 1877 and the subsequent reoccupation of Eastern Turkestan by the Chinese forces did not result in any adverse impact on the Indo-Central Asian trade. Several factors were responsible for the continuance of this trade even after Xinjiang had become a Chinese possession in 1878. First the Chinese authorities did not interfere in the long established trade between India and the southern oasis-towns of Xinjiang. Their restrictions on the import of Indian tea and opium also proved ineffective in the face of the smuggling of these commodities in connivance with the Chinese customs officials. Second, Xinjiang did not possess an industrial base to cater for its demand for manufactured and other goods that were imported mainly from Russia through the Kokand–Kashghar route and partly from India through the Srinagar–Leh–Yarkand route. Third, Indian trade with Bukhara, Kokand and other parts of Russian Central Asia, which was hitherto carried through the Peshawar–Kabul–Bukhara route, was partly diverted to the Srinagar–Leh–Yarkand route after the Russians imposed strict restrictions on the import of British Indian goods into Russian Central Asia. Fourth, the arrival at Leh of Central Asian pilgrims en route to their Haj pilgrimage to Mecca via India also contributed to the increased Central Asian imports into Ladakh as they brought along with them a large quantity of bullion, silks and ponies for sale in Leh or Srinagar. Fifth, the complete security of the Srinagar-Leh Yarkand route which was devoid of robbers and the facilities available at various stages of this route as far as the supply of ponies, transporters, food and fodder was concerned, also helped in the development of Ladakh as an important trading centre.

After the conclusion of the Russo-Chinese treaty at St. Petersburg in 1881 the bilateral overland trade between Xinjiang and Russia scaled new heights. This in turn affected Indian trade with Xinjiang in more than one way, as it had now to face stiff competition from Russia in that quarter. The import of Khotanese gold dust and Chinese silver ingots into Ladakh was now substituted by that of Russian gold and paper rubles. The Yarkandi felts and Khotanese carpets became dearer in Ladakh and Kashmir as their prices appreciated due to increased demand in the Russian markets. A lesser quantity of Turfani fine wool was imported into Ladakh, as the same was being supplied to Russian Turkestan in large quantities. Import of coarse wool and pashm from Chinese Turkestan dropped considerably in value from 1930 onwards. But the prosperity of the Russo-Xinjiang trade also meant a corresponding increase in Indian trade with Xinjiang. Since Xinjiang's exports to Russia far exceeded the imports from that end, the balance was met by ruble payments which ensured a regular and abundant supply of Russian currency in the markets of Kashgharia. This in turn led to increased imports from India which were now paid back largely in Russian gold and paper rubles and raw silk. However, hemp drug (*charas*) continued to be imported into Ladakh despite its heavy taxation in India. An average of 3,500 maunds of *charas* was imported annually into Ladakh from Chinese Turkestan during the years 1882 to 1932. Most of this *charas* was consumed in Punjab and Oudh whence it was carried by Indian traders via *the* Leh–Kullu route.

The years 1890 to 1908 saw an unusual increase in coral exports to Chinese Turkestan from India. During this period coral worth about twenty four lakh rupees was exported. There existed a strong tradition among the Central Asians to use coral ornaments. About half of the coral exported to Chinese Turkestan was re-exported from there to Russian Turkestan. But this trade, which was entirely in the hands of Indian merchants, reached its lowest point in 1918 when coral worth about 180 rupees only was exported from Ladakh.[30] At the close of the nineteenth century Russian currency began to occupy an important place in the Indian imports from Chinese Turkestan. Ruble import was resorted to by the Indian traders as a forced medium for sending back their sale proceeds to India. Russian gold and paper rubles worth about 80 lakh rupees were imported into India via Ladakh during the years 1897 to 1919.

With the establishment of Soviet power in Central Asia, bilateral trade between Xinjiang and Russian Turkestan received a setback. It was now diverted to Ladakh and Kashmir. This resulted in an unusual increase in the Indian exports to Xinjiang through Ladakh. During the years 1917 to 1931 goods worth about 285 lakh rupees were exported from India to Chinese Turkestan through Kashmir and Ladakh, of which European cotton goods and silks, dyed skins, indigo, spices and tea constituted the main items. In the same period, merchandise valued at about 330 lakh rupees was imported from Chinese Turkestan into Ladakh and Kashmir of which *charas*, Russian gold and paper rubles, gold dust, raw silk, carpets and felts were the main items. The balance of trade thus stood in Xinjiang's favour. The Indo-Central Asian trade through Kashmir and Ladakh, which had scaled an unprecedented height of more than 68 lakh rupees during the year 1920–21,[31] began to decline in value soon after the opening of Russian overland trade with Xinjiang. From the late 1930s onwards, internal disorders in Xinjiang began to act as a stumbling block to this trade which finally ceased to flow after 1949 following the Communist takeover of Xinjiang and its subsequent closure to outside traffic. This deprived Ladakh and Kashmir of their importance as emporiums of the Indo-Central Asian overland trade that passed through a network of caravan trails criss-crossing Ladakh and Kashmir.

Impact of trade on society and culture

The passage of lucrative trade between India and Central Asia through Ladakh and Kashmir left a distinct impression upon the local society and economy. Leh, the capital of Ladakh, acquired a cosmopolitan character during the trading season when merchants of different nationalities coming from Bukhara, Andijan, Kashghar, Yarkand, Kabul, Badakhshan, Tibet, Amritsar, Hoshiarpur, Kullu, Nurpur, Bushahr and Kashmir arrived at Ladakh to sell their goods. The sparsely populated town of Leh suddenly hummed with activity which lasted throughout the summer months. This trade brought economic prosperity to all those who were in one way or the other involved in it. However, Ladakhis benefited from this trade only partially because very few of them carried on their own business. Most of the Ladakhis being poor and illiterate contented themselves to act as pony-drivers, porters, suppliers of pack animals and forage to traders. Those Ladakhis, Arghuns and Kashmiri settlers in Leh who traded with Central Asia and Tibet turned into a richer and affluent class as compared to the majority of Ladakh's population which was poor. Buddhist *Skudrags* of Ladakh were generally engaged in domestic trade. Opposite to this, the Muslim traders of Leh were involved in trade on all routes and they maintained family and trade links in Skardu, Kashmir, Lahaul, Nurpur, Yarkand and Tibet. They wielded considerable influence in Ladakh which was quite disproportionate to their smaller numbers.

Due to the circumstances of trade a number of Central Asian, Hoshiarpuri and Kashmiri merchants were obliged to take up their abode in Leh. Most of these traders indulged into marriages of convenience with the local Ladakhi girls and became an inseparable part of the Ladakhi society. Among them were the Kashmiri Muslims who traded with Tibet and exported pashmina wool and tea to Kashmir. Similarly some Central Asian merchants would stay back in Ladakh in times of political uncertainty in their homeland or owing to the closure of passes, and take temporary Ladakhi wives. The result was that a new hybrid class of *Arghuns* came into being. They were the offspring of Kashmiri or Central Asian traders who had taken Ladakhi wives, temporary or permanent. By and large the *Arghuns* continued with the tradition of trading with Central Asia or Tibet. Some of them also acted as the carriers of merchandise that belonged to traders.

The passage of trade through Ladakh and Kashmir influenced the dress, food and drinking habits of its people. Machine-made cloth brought by Indian traders, velvets, otter skin hats, Khotanese or Chinese silk fabrics and Yarkandi cotton cloths were increasingly used by the local people, particularly the elite classes, as their dress material. Chinese brick-tea imported from

Lhasa and Yarkand was largely consumed within Ladakh. Part of this tea was also sent to Kashmir. Similarly, Yarkandi *pilau* (rice cooked with meat and spices) became an important addition to the Ladakhi cuisine. The use of steamed dumplings of meat locally called *mo·mo* also became common in Ladakh. The affluent sections of Ladakhi society used to wear long leather boots that were imported from Yarkand. Some items of Yarkandi dress such as *pichak tungyu* (conch handled knife) and *kosa masi* (long calf leather footwear with a separate pair of leather slippers) became a fancy item of Ladakhi dress. The use of Chinese teacups, jade articles from Khotan and Lhasa, Yarkandi steamers and teapots was also introduced in Ladakh and Kashmir. Central Asian visitors and *Arghuns* who frequently travelled between Xinjiang and Ladakh, introduced in Leh certain dance forms such as tall-man, dragon, lion and boat (*kishti*) which were prevalent in Xinjiang.[32]

The farmers in Leh and surrounding villages like Stok, Shey, Spitok, Chushot, Choglamsar and other villages situated along the trade routes took to extensive cultivation of lucerne grass. The large-scale introduction of a perennial variety of this grass called Yarkandi ol is obviously the direct result of Ladakh's commercial intercourse with Yarkand.[33] The agriculturists profited by leasing out their grass fields to traders and pony drivers to feed their ponies on. But this trade put a drain on the food resources of Ladakh. During the Dogra rule an average quantity of three to four thousand maunds of barley grain was collected as land revenue from Ladakhi peasants.[34] It amounted to less than half of the actual requirements of the traders.[35] The deficit was met partly by forced procurement of barley from Ladakhi cultivators at much lower prices than the actual market rates.[36] The result was that Ladakhi agriculturists faced immense hardships as they were also required to deliver the grains at the godowns. State granaries were located at Leh, Saspol and Lamayuru, where barley was supplied to traders at fair prices.

Ladakhis indulged in large scale weaving of coarse woollen cloth and blankets. They also manufactured sacking which was in constant demand from the traders who required it for packing their goods. According to Cunningham, the quantity of blanketing and sacking annually consumed in Ladakh in connection with the carrying trade amounted to about 120,000 yards of one foot width in the manufacture of which about six and a half lakh lbs. of wool were consumed.[37] But for this home-made wool, Ladakhis were almost wholly dependent upon imports for their day to day requirements such as manufactured cotton cloths and silks, tea, spices, utensils, leather goods, felts, carpets etc. Since these commodities were abundantly available at cheaper prices in the Leh market, no incentive was thereby left to promote the local industry.

The continuation of the traditional trade relationship between Ladakh and Tibet under the Dogras was matched by the strengthening of cultural ties between the people of Ladakh and Tibet. Apart from the frequent exchanges of Buddhist monks, this trade resulted in the settlement of some Ladakhi

Muslim trading families in Tibet. The Kashmiris, Baltis, Ladakhis and Tibetans relished the same food and drinking habits. Brick-tea, *sattu* and apricots formed the favourite diet of these people. Similarly they used the same articles like saffron, *nabat* and yak tails for worship and as religious offerings.

Geographical proximity and cultural affinity in terms of religion, customs, dress, food and drinking habits provided a strong base for the continuation of the socio-cultural links between the peoples of Central Asia and Kashmir even after the two regions had been incorporated in the Chinese, Russian and British empires. Even today both the Central Asians and Kashmiris drink the same hot tea in the same type of *piala* (a big porcelain cup) and relish the same *pilau* (stuffed rice with meat). The people in both the regions are rice-eaters and non-vegetarian. The *samovar*, which stands for a tea-urn in both the Russian and Kashmiri languages, continues to be a popular institution among the peoples of this whole region. Indian tea, spices and other manufactured goods like muslins, shawls and brocades have been popular among the Central Asians. The musical instruments used in Kashmir like the *santoor*, *surnai*, *rabab*, *sarangi* and *tumbaknari* (the single membrane drum) are virtually the same as those used in Central Asia. The abundance of apples, grapes, melons, walnuts, mulberries, apricots and other delicious fruits in Central Asia has a parallel in the luxuriant and variegated horticultural produce of Kashmir. Kashmiri families bearing surnames like *Akhoon*, *Beg*, *Kashghari*, *Turki*, *Bukhari*, *Nakshbandi*, *Gani*, *Mirza*, *Mughal*, etc., are the living examples of the cultural assimilation that took place as a result of the Central Asian immigration into Kashmir in medieval times.

Apart from these sociological and ecological factors, the movement of men (e.g. pilgrims) and materials (books, Kashmir shawls, gold coins, Central Asian silk, Indian tea and spices, coral etc), formed an essential ingredient of the socio-economic intercourse between Central Asia and Kashmir until the early twentieth century. Thus Kashmir and Ladakh played a signficant role in India's relations with Central Asia on account of their geographical proximity and cultural affinity with the region. As a result of its linkage with the Silk Route, Kashmir and particularly its frontier territories of Ladakh, Baltistan, Gilgit and Hunza became a major gateway in the Indo-Central Asian exchange of men, materials and ideas. The caravan traders acted as a vehicle of this cultural interaction. During the nineteenth century the Anglo-Russian rivalry over Central Asia lent new dimensions to this relationship. As contacts with Russian parts of Central Asia were subjected to several inhibiting influences, the relations with Xinjiang, particularly trade, scaled new heights. But when Xinjiang became part of Communist China in 1949, which was followed by the closure of Indian Consulate General in Kashghar and sealing of Xinjiang's borders, all traditional links between Ladakh and Xinjiang were disrupted.

Conclusion

As the traditional and historical trade links between India and Central Asia via Kashmir and Ladakh ceased to be functional after 1949, time has come to reflect upon the legacy of such a rich and productive experience and revive these links and trade relations on a new and dynamic footing. Chinese experience in Xinjiang shows how this remote north-western region has been brought closer to China's mainland both by air, rail and road network, notwithstanding the enormous distance and inhospitable deserts intervening between the two regions.

China and the Central Asian Republics of Uzbekistan, Kazakhstan and Kyrgyzstan are extending their transportation links via Osh, Naryn (Turgart pass/Irkeshtam) through Kashghar and Yarkand in Xinjiang to be linked to the Karakoram Highway running through Pak-occupied Kashmir, so that these Central Asian countries secure access to the sea via Pakistan. There is also a proposal to link cis-Pamir mountain territories of Karategin (Komsomolabad, Garm, Tajikabad, Jirghital), Darvaz and Vahio (Tavildara) in eastern Tajikistan, to Kyrgyzstan and the Kashghar region of Xinjiang by formation of a transport and economic corridor through Suhov and Kyzyl Su (in Kyrgyzstan), Sary Tash in Pamirs, and via Irkeshtam to Kashghar in China. This 750 km route is reported to have no high passes or inaccessible sections and could be made operational throughout the year. Besides, there already exists a road on some sections of this route and the whole section from Dushanbe to Kashghar (750–800 km) has asphalt or pebble stone cover. Only a few sections of this road are reported to be not metalled.

This transport network is sought to facilitate development of marginal, peripheral and developed areas of Tajikistan, Kyrgyzstan and southern Xinjiang. It will embrace the Pamir tract, the Karakoram Highway and new routes to Kazakhstan and Kyrgyzstan (through the Turgart pass). The Euro-Asian highway project has envisaged the completion of the following transborder roads, to connect the Central Asian Republics and the Xinjiang region of China:

- Khorgos (Kazakhstan)–Kuldja (Xinjiang)
- Bishkek–Turgart Pass (Kyrgyzstan)–Kashghar (Xinjiang)
- Osh-Sary Tash (Kyrgyzstan)–Kashghar (Xinjiang)

It is high time that the following steps are initiated to revive the traditional India–Central Asia overland routes through Ladakh on a priority basis:

- Whereas the recent reopening of border trade at Nathu La in Sikkim is a welcome development, the traditional India–Central Asia overland trade routes via Leh, Yarkand, Kashghar and onwards to the Central Asian Republics, and Leh–Demchok–Gartok–Lhasa need to be reopened in the same manner.

- The proposal of the Jammu and Kashmir Government to open the Leh–Demchok route to western Tibet as a viable and easier alternative route for pilgrimage to the Kailash–Mansarovar across the Line of Control (LoC) in Ladakh needs to be pursued and to be accepted by the Chinese. This will help in reducing the journey time and provide a safer passage to pilgrims. (More than 300 pilgrims died due to landslides at Malpa in the Uttarakhand hills in October 1998).
- The possibility of opening an oil/gas pipeline linkage across the LoC in Ladakh with the proposed Xinjiang–Kazakhstan pipeline, needs to be studied.

2 'Great Game' on the Kashmir frontiers

K. Warikoo

Introduction

The Anglo-Russian agreement of 1873, whereby the British and Russian spheres of influence in Central Asia and Afghanistan were mutually agreed upon, instead of ushering in a new era of cordial relations between the two rival powers added new dimensions to the 'Great Game'. Whereas this agreement in effect gave the two sides freedom and a sort of legitimacy to their advance within their respective zones,[1] at the same time it brought to the surface the new problem of the actual delimitation of Afghan, Chinese and Russian frontiers in the upper Oxus region of the Pamirs. British attention was drawn to the complexity of this question by British officers like Gordon, Trotter and Biddulph who in 1874 explored the Wakhan and Pamirs area. They discovered that the Afghan territory in the eastern extremity lay on both sides of the river Oxus, which under the 1873 agreement was declared to be the dividing line between Afghanistan and Russia. This discovery disputed the very foundation of this accord. On examination of the Hindu Kush passes, the British explorers found them easy to cross, thus making India vulnerable to attack from across the Hindu Kush. Both these discoveries were important from the strategic point of view and the British modified their frontier policy accordingly. The deputation of Biddulph in 1876 to survey the Hindu Kush passes, which was followed by the establishment of a British agency in Gilgit under the same officer in 1877, reflected the new British strategy to meet the challenge posed by the Russian approach to the Pamirs.

C. M. MacGregor, the Quarter Master General of the Indian army (1880–85) also contributed a lot to mould the British policy in the cast of Russophobia. In his book entitled *Defence of India*, which was published in 1884, MacGregor openly aired his anti-Russian sentiments and recommended a number of measures designed to achieve the fragmentation of the Russian empire. He apprehended a Russian attack on India from the direction of Kabul, Herat, Chitral and Gilgit. MacGregor wanted the British government to:

- play the game of diplomacy with Russia;
- destroy Russian trade;

- form a coalition with Austria, Germany and Turkey;
- despatch commissions for demarcation of the northern and north-western frontiers of Afghanistan;
- transfer Herat province to the British;
- transfer to the British the regions of Wakhan, Chitral and Yasin;
- attract Persia to the British side;
- forge closer links with China;
- improve means of acquiring information about the Russian movements;
- despatch emissaries for instigating disorder in the Khanates of Central Asia and among the Turkmens; and
- fragment the Russian empire.[2]

When his book, though a classified publication, fell into the hands of the Russian Military Attaché at the Russian Embassy in London, a serious view was taken of the British plans to fragment the Russian empire.[3] MacGregor's proposals aroused Russian suspicions about the British intentions and became the subject of diplomatic exchanges between the Russian and British Governments.[4]

Meanwhile in 1885–86, Ney Elias was deputed by the British Indian Government to reconnoitre the frontier areas of Sinkiang, Wakhan, Badakhshan and the Oxus headwaters. On his return Elias came up with some startling information. He pointed out that the two extreme ends of the Afghan and Chinese frontiers in the Pamir area could be stretched and joined together to form a neutral buffer, thus preventing any direct contact between India and Russia in the Hindu Kush and Dardistan region. Elias's discovery was to become later 'a cardinal feature of British policy towards the Pamirs'.[5] The Wakhan corridor stands there even today as a testimony to Elias's stratagem. The British forward policy in the area lying south of the Hindu Kush was also active due to reports of direct dealings between the Hunza Chief, Safdar Ali, and a Russian Officer, Grombchevsky. To meet any possible Russian threat the British adopted a two-pronged strategy. They started inducing the Chinese to occupy as much area as possible in the Pamirs in order to keep the Russians at a safe distance. To achieve this end Francis Younghusband, who had previously been despatched to survey the Khunjerab and Karakoram ranges in 1889, was deputed in June 1890 to the Pamirs to see 'where the Afghan and Chinese boundaries should be made to meet'.[6] Younghusband did succeed in persuading the Chinese authorities in Sinkiang to send a force to occupy Somatash, but he found himself outmanoeuvred by the Russians who had already occupied Somatash, thanks to the groundwork done by the Russian Consul in Kashghar with the Chinese *Taotai* there. A timely hint from the *Taotai* had enabled Petrovsky to forewarn Grombchevsky, the Russian frontier officer at Marghilan, about the British ambitions. In fact Younghusband's mission to Chinese Turkestan 'activized the Tsarist government which for some time had been following a policy of "wait and see" towards the Pamirs which resulted in the visit of the Governor General of

Turkestan, A. Vrevsky, to the Alai Pamirs'.[7] When Younghusband was making the return journey to India via the Pamirs in August 1891, he was forcibly expelled by a party of Russian Cossacks led by Colonel Ivanov near Bozai Gumbaz which was declared to be within Russian territory. This fresh development sparked off a crisis in the Anglo-Russian relations but conflict was averted with the Russians adopting a low profile. However, this incident made the British occupation of the Dardic territory lying below the Hindu Kush an urgent necessity.

By now the strategic importance of the Pamirs had become clear to both the British and the Russians. The Pamirs was the meeting point of the Kashmir frontiers in Gilgit, Hunza and Chitral, the Afghan provinces of Badakhshan and Wakhan, the Russian territory of Kokand and the Sarikol area of Chinese Turkestan. It was a potential gateway to India. The British strategy geared itself to the task of creating a barrier between Russian and British empires right on the Pamirs, simultaneously extending their effective control over the frontier areas in Gilgit, Hunza, Chitral and Yasin through the Maharaja of Kashmir. Several factors contributed to the shifting of British interest to Kashmir frontiers which from the late 1870s onwards became the focal point of the 'Great Game'. First, by occupying Kokand, Russia had acquired a legitimate claim over the Pamirs which were the summer pastures of Kirghiz subjects of the erstwhile Khanate of Kokand. Second, Chinese Turkestan now came within the effective range of Russia from their newly acquired territory in Kokand. As the Chinese authorities in Sinkiang were weak at that time, the British feared that the Russians would occupy Sinkiang next which would then prove to be an important supply base in the event of any further Russian attack on India via the Kashgar–Karakoram route. Third, the reports of Gordon, Trotter and Biddulph had underlined the strategic importance of the Hindu Kush passes which were now considered to be easily accessible, thereby making India vulnerable to any outside attack from across the Hindu Kush and the Pamirs. Fourth, neither the Chinese nor the Afghans possessed any effective control over the Pamirs, leaving the area open for Russian penetration. Finally, the reports of Russian officers having explored the Pamirs and the Hindu Kush region further strengthened the British apprehensions.

Confronted with these strategic considerations, the British encouraged the Maharaja of Kashmir in his adventurist propensities to bring the warlike and unruly Dardic tribal chiefs inhabiting the obscure mountainous valleys of the Hindu Kush and Karakoram under his effective control. The Maharaja was given the freedom to choose any means from conciliation to military expeditions or both and it was soon found that the Maharaja could not stop Dardistan from drifting into a state of turmoil and instability. This fluid situation was the result of frequent internecine wars between the tribal chiefs and also due to their occasional attacks on the Kashmir troops which often resulted in the lapse of Kashmir's control over these frontier dependencies. The wavering loyalty of the Muslim frontier chiefs towards the Hindu ruler of Kashmir and through him to British India; the exalted image of Tsarist

Russia in this region; the display of active Russian interest in this frontier belt; and the open defiance of British power by Safdar Ali, the Chief of Hunza, who claimed both the Russian and Chinese support, all forced the British to resort to direct action. The idea behind several military expeditions carried out jointly by the Kashmir forces and the British officers against Hunza, Nagar, Chitral and Yasin was to put the defence of north and north-western frontiers on a firm footing. Once these unruly tribes were coerced into subjugation to Kashmir, a strong military garrison was established at Gilgit, thus making it the nucleus of the defence arrangements. Besides, a new force was raised from amongst the local tribes, which served a two-fold purpose. First, it channelized the energies of warlike and turbulent Dardic tribals usefully and they now became an inalienable part of the defence of the north-western frontier. Secondly, it reduced the actual expenditure of the British Indian Government which was already spending lakhs of rupees for maintaining a garrison at Gilgit.

British diplomacy achieved the second objective of creating a buffer in the Pamirs between the British and Russian empires by concluding the Pamirs Boundary Agreement in 1895. Although the 'Great Game' appeared to be over now, the focus of the two-power rivalry shifted to Sarikol and Taghdumbash Pamirs where Chinese possessions in Sinkiang and the British Indian territory of Hunza converged. Russia's interest in this area emanated from its fears about the possibility of the British extending their control over the Taghdumbash Pamirs where Chinese authority was minimal. The British and Russian official representatives kept themselves engaged in a war of nerves, each trying to outwit the other to have closer access to the Chinese authorities in Sinkiang. The British support to certain Hunza claims on Raskam lands and Taghdumbash Pamirs attracted strong Russian reaction. On their part the British considered the establishment of Russian post in Tashkurghan with the consent of the Chinese as a new move to threaten Hunza and Gilgit and from there, Kashmir and India. So far as the area around the Kashmir frontiers was concerned, the Anglo-Russian convention of 1907 did not end their mutual suspicions in this region. This is borne out by the fact that a number of Russian military officers came to explore this frontier even after 1907. It was only during World War I that the two sides actually acted in unison to meet the common threat from Germany. The close cooperation between the British and Russian Consuls at Kashghar during this phase in hounding out the suspected German agents from the Pamirs was natural in these circumstances. In the following pages, we shall discuss various strands of the 'Great Game' which concerned the north and north-western frontiers of Kashmir, before the October 1917 revolution in Soviet Russia.

Extending British control up to the Kashmir frontiers

Mayo's Viceroyalty was marked by a distinct shift from the policy of non-interference in the affairs of Kashmir as laid down by his predecessor in 1868.

He initiated a process of enforcing British influence over all the external diplomatic proceedings of the Kashmir ruler. Though neither Mayo nor his foreign department questioned Kashmir's jurisdiction over Gilgit, they sought to curb the Maharaja's propensities for further expansion in that direction, particularly when they learnt about the exchange of correspondence between the Maharaja of Kashmir and Mir of Badakhshan on the subject of Kashmir's tutelage over Punyal and Yasin.[8] The British were not totally unaware of the state of affairs in Gilgit and adjacent chiefships within Dardistan, thanks to the extensive field work by Frederick Drew and Leitner in that area during the 1860s. But their studies did not provide any insight into the strategic details of this area in the context of Russian approach to the Pamirs. Even Pandit Munphool, a native political agent despatched on a probing mission to the region in 1865, could not bring in all the desired information. It was Mayo who sought to fill the existing gaps in his stock of strategic information about Dardistan and adjacent areas, by deputing a team of trained native explorers in the train of Forsyth's first mission to Kashgar in 1870. By this time another British explorer, Hayward, had started exploring the region under the patronage of the Royal Geographical Society, London and with the knowledge of the Indian Government. By mid-1870, Hayward had completed his two visits to Gilgit and Yasin and during these visits he did not keep secret his political proclivities against the Kashmir ruler. However, he did keep the Government of India posted about what he heard and saw and also about what transpired during his meetings with the local tribal chiefs. Partly influenced by Hayward's description of alleged atrocities by Kashmir troops in Yasin and partly motivated by a desire to put a stop to Maharaja Ranbir Singh's forward policy, Mayo decided to restrain the Kashmir ruler from extending his authority any further.[9] At a meeting in Sialkot he asked the Maharaja to 'communicate with the British government on matters of importance more frequently than heretofore'.[10] On his part, Ranbir Singh stuck to his stand that the frontier territories in Ladakh, Baltistan, Gilgit, Yasin, Punyal, Hunza and Nagar formed a part of Gulab Singh's territory long before the Amritsar treaty was signed in 1846.[11] However, the British policy enunciated at Sialkot was not destined to stand the test of time. The circumstances of Hayward's murder at the hands of the Mir of Yasin in July 1870 must have made the British alive to the dangers implicit in any direct dealings with the turbulent chiefs in Dardistan. Besides, the imperial requirements that demanded the extension of British influence up to the Hindu Kush watershed for offsetting any future Russian advance from their frontier outpost at Osh could not be met by Mayo's policy of restricting the Kashmir ruler's forward movements. In 1874–75, the British Indian Government received valuable survey reports on the Wakhan, Pamirs and Sarikol region from Colonel Gordon, Biddulph and Trotter, who had been purposefully sent to Kashgar in the company of Forsyth's second mission of 1873. Gordon's disclosures about the existence of a practicable road from the Russian military post of Osh across the Alai to Sarikol brought into sharp focus the strategic importance of this area. When Gordon pointed to the vulnerability of India from the direction of Wakhan via the Ishkoman and Baroghil passes, the need to strengthen the British position in the tribal territories of Gilgit, Yasin and Chitral assumed importance. The discoveries made by Biddulph during his exploration in the Wakhan area during 1874 and of the passes lying south of the Hindu Kush during 1876, only reinforced Gordon's viewpoint. By 1876, the British Indian Government appeared to have been convinced about the necessity of extending Kashmir's control over Chitral and Yasin right up to the south of the Ishkoman and Baroghil passes. The incorporation of Kokand by Russia, which pushed the Russian frontier beyond Osh, only helped in catalysing British counter-moves in Dardistan and Kashmir. It was against this background that Mayo's Sialkot stand of 1870 was replaced by Lytton's policy enunciated at Madhopore in 1876.

The stationing of a British agent in Gilgit, 1877

The importance of Gilgit as a convenient base for extending the British influence up to the territories lying south of the Hindu Kush had now been fully realized. To quote Alder, it was a 'natural choice' being situated at the 'hub of routes leading off to all parts of Dardistan'.[12] Though Eastern Turkestan continued to be regarded as a rich supply base to support any Russian force coming from its western flanks, the defence of the Hindu Kush watershed was of immediate concern to the British. Following the advice of his predecessor (Lord Northbrook), Lytton formulated his proposals about the future course of action to be taken in this frontier belt, which he finally conveyed to Maharaja Ranbir Singh personally at Madhopore on 17 and 18 November 1876. Lytton impressed upon the Kashmir ruler the need to strengthen Indian frontiers by assuming control over the territory that lay between the Hindu Kush and Kashmir frontier, in order to secure command of such passes as were thought to be practicable for the passage of Russian forces.[13] To the British, it was now of vital importance that the states like Chitral and Yasin 'should come under the control of a friend and ally' like the Maharaja of Kashmir, 'rather than be absorbed by powers inimical to Kashmir'.[14] But while encouraging Ranbir Singh to obtain 'an effective but peaceful control over the countries lying between those passes and the Kashmir frontier namely Chitral, Mastuj, Yasin and their dependencies'[15] the British secured the right to station an agent at Gilgit 'to collect information regarding the frontier and the progress of events beyond it'.[16] The Kashmir ruler relented to this measure only after obtaining written assurance from Lytton that the Gilgit Agent would in no case interfere in his internal administration. The Maharaja also volunteered to connect Gilgit, Srinagar and Jammu with the British Indian telegraph system. After having obtained Kashmir's concurrence, the formal announcement for the appointment of Captain J. Biddulph as Officer on Special Duty in Gilgit was made on 22 September 1877.

The tasks assigned to him were not only limited to the mere collection of information about the topography and resources of the territory beyond the Kashmir frontier, but also to the extension of British influence among the tribal people by cultivating friendship with them.[17] Biddulph was joined soon after by a medical officer who was to help in popularizing the British image among these people through the healing touch of a doctor. Though Biddulph was quite successful in keeping a watchful eye on the Russian movements in Badakhshan, Afghan Turkestan and Kokand, he could not win over the support of Kashmir officials. One cannot dismiss his accusations against the state officials of plundering his baggage and also of creating a wedge between the tribal chiefs and the British Government as unfounded. Biddulph's suspicions were strengthened by the experience of Shah Khushwakt, a native agent sent by him in May 1878 to Hunza, Kabul, Bukhara and Kokand on a spying mission,[18] who suffered imprisonment in Hunza during his return journey for being a British emissary. He believed that no person other than Babu Nilambar, one of the Kashmir ministers, had disclosed to Ghazan Khan of Hunza through his Vakil information about the nature of Khushwakt's activities.[19] To Biddulph's surprise, five thousand rifles gifted by the Government of India to the Kashmir ruler for strengthening his frontier defence, had not been sent to Gilgit.[20] And when Henvey, the Officer on special duty in Kashmir, raised this issue with the Kashmir Durbar, he was curtly told that no conditions had been attached to the gift of these arms.[21] Obviously Kashmir disparaged the foisting of a British agent on its territorial jurisdiction. By getting the Kashmir Governor at Gilgit removed, Biddulph only added to his difficulties. The end result was that the British did not achieve the desired objective of creating direct influence among the tribal chiefs. Nor could they enforce their authority over Kashmir, which continued to resist such interference.[22]

The Gilgit Officer, however, continued independently in this remote end of the Indian northern frontier up to 1880. The matter came to head on 28 October 1880 when the successful assault by Hunza and Yasin on Gakuch and Sher Kila exposed the military weakness of Kashmir to hold this territory. The precarious situation on the Gilgit frontiers had even put the personal safety of the British agent at Gilgit in danger. The inaccessibility of Gilgit which was poorly connected with Srinagar by a rough track running over high mountains for 230 miles (370 km) and open to traffic for six months only, had encouraged the frontier chiefs of Dardistan to be irresponsible. The frontier uprising of late 1880 proved to be the proverbial last straw for the Gilgit Agency. To Ripon, who already considered the Gilgit appointment as 'a mistake'[23] the advantages accruing from its continuation were disproportionate to the embarrassments and anxieties suffered by the British. Finally it was decided in July 1881 to withdraw the Gilgit Agent, which, however, did not mean British non-involvement in this area. In fact, the Agency was only kept in abeyance until 1888. The British reserved their right to reappoint the Officer at Gilgit when necessary. The Secretary of State, while consenting to this measure, gave vent to his reservations over the issue by underlining the point that the withdrawal of the Agent might practically close a valuable channel of information as to the course of events in the countries between Kashmir and Russian Turkestan.[24]

Re-establishment of Gilgit Agency, 1889

The British policy of withdrawing the Gilgit Agency in 1881 was bound to change in the face of intricate problems involved in the defence of the northern frontiers. Their desire to have a firm control over the territories up to the Hindu Kush could not be met by being absent. The Afghan pressures on Chitral, which had already been drawn close to the Indian empire through its allegiance to Kashmir in late 1876, assumed seriousness in 1882 when Kabul claimed this territory 'as one of its protected states'.[25] Though the Government of India was quick in rebutting such claims,[26] the reported Russian intrigues in and around Chitral caused additional concern. The newly-established Resident in Kashmir could hardly extend effective control of the Durbar or even the British Indian government over the turbulent chiefships of Hunza and Nagar. The Panjdeh[27] crisis of 1885 also helped in focusing the British military opinion on the need to make elaborate arrangements for the defence of the north-west frontier. So it was not a mere coincidence that the year 1885 was marked by the despatch of two exploratory missions, one under Col. Lockhart, the Deputy Quarter Master General of India, to survey the lands south of Hindu Kush via Gilgit and Chitral, and the other under Ney Elias to Chinese Turkestan and the Pamirs. Meanwhile the Defence Committee's recommendation for extending the Srinagar–Rawalpindi cart-road to Gilgit and Chitral,[28] had been approved by Dufferin. The new Commander-in-Chief, F. Roberts, even suggested that 'we should have political control over the country around Chitral and Gilgit in order to secure the approaches to the former by the Dora pass and to the latter through Wakhan'.[29] The purpose of Lockhart's mission was to establish friendly relations with Chitral and Kafiristan and also to obtain accurate information about the routes, passes and resources of the country lying below the Hindu Kush.[30] During 1885–86, he not only surveyed about twelve thousand square miles of territory of Gilgit, Chitral, Hunza, Taghdumbash Pamirs and all the important passes across the Hindu Kush,[31] but also secured the Chitral ruler Aman-ul-Mulk's promise to allow the passage of British troops through his territory in the event of a Russian attack from that side.[32]

Lockhart's recommendations, though not accepted at the time, did provide a basis for the future British policy towards this frontier belt. He wanted the British Government to acquire Gilgit from the Kashmir Durbar and establish there a garrison of locally raised troops under a British commandant carrying out both the political and civil functions.[33] The importance of Gilgit 'as the defensive nucleus of Dardistan'[34] was once again underlined by Lockhart in the following words.

> The acquisition of Gilgit would secure us the continued loyalty of Chitral, carrying with it our right of way through the Mehtar's dominions, and

> his active cooperation in time of need. In my opinion it would ensure the safety of Hindu Kush.[35]

By May 1887, the Secretary in the Foreign Department (GOI) H. M. Durand, had reached almost the same conclusions. By recommending the re-establishment of the Gilgit Agency with a garrison of Kashmir troops and local levies, Durand wanted the British Indian Government to adopt a more active policy towards this northern frontier so that in the event of any difficulties with Russia, Kashmir would not be 'more or less shaky and inclined to hedge'.[36] It was in October 1887 that more clear directives reached the Resident in Kashmir regarding the revised policy of the Indian Government about the frontier defence. He was informed:

> [the] Time has come for establishing on the north-west frontier of Kashmir an effective political control, which will enable us to watch the passes of the Hindu Kush and the country beyond, and a military organization sufficient both to control the Chiefships over the border and also to check, in the event of war with Russia, any demonstration towards the passes not backed by a respectable force.[37]

Since the British wanted to execute their scheme with the cooperation of the Kashmir Durbar, the British Resident in Kashmir, Plowden, was asked to use his influence there to make it smooth running.[38] In fact, Plowden was authorized to promise a military rank to the Maharaja's younger brother for making the Gilgit scheme palatable to the Kashmir Durbar.[39] But Durand's guidelines to the Resident left little room for doubting the British intention to secure political control of the frontier territory in Gilgit and up to the Hindu Kush, though the Maharaja's Governor and his troops stationed in Gilgit were to be nominally left under the Kashmir Durbar.[40]

The need to strengthen Kashmir's hold over its frontier territories assumed urgency in view of a joint attack by Hunza and Nagar on Kashmir posts at Chaprot and Chalt in early 1888. Besides, the Indian Government had in mind Ghazan Khan's refusal to permit Lockhart through his territories in April 1886 and also the Chinese links with Hunza. It was against this backdrop that Dufferin decided to depute Capt. A. G. A. Durand (younger brother of the Foreign Department Secretary, H. M. Durand) on a mission to Gilgit. He was required to:

> report on the military position at Gilgit with reference to the recent tribal disturbances and to future possible complications with Russia, and to work out a scheme for rendering Gilgit secure without the aid of British troops and for dominating from Gilgit, through the Kashmir forces, the country up to the Hindu Kush; thus rendering Kashmir territory thoroughly secure against attack and guarding against the possibility of a Russian force penetrating to Chitral and threatening our lines of communication between Kabul and Peshawar through the Kunar Valley.[41]

On his return in late 1888, Capt. Durand while reporting about the visit of a Russian officer (Grombchevsky) to Hunza in the autumn of that year, underlined the need to close the gap between the Chinese frontier post at Aktash and the Afghan frontier at Lake Victoria, which otherwise could give the Russians access to Hunza.[42] Taking full note of the possible lines of Russian advance to Chitral and Hunza, Durand sought to seal them by proposing the re-establishment of a political agency at Gilgit and the stationing of about 2,000 Kashmir troops there to function under four British Officers.[43] He also recommended the improvement of the Kashmir–Gilgit–Chitral road and the early completion of the telegraph line to Gilgit.[44] The local chiefs of Chitral, Punyal, Hunza and Nagar were proposed to be encouraged to become an inseparable part of the Indian empire, by granting increased subsidies to them.[45]

Owing to Dufferin's departure from India, the task of implementing his active policy towards Gilgit fell upon his successor Lansdowne who readily accepted Durand's scheme. Lansdowne's task was rendered easier by the political changes in Kashmir in April 1889 when the state administration was brought under the direct control of the Indian Government through the Resident. Accordingly Capt. Durand, along with two British officers, Dr. Robertson and Lt. Manners-Smith, arrived in Kashmir in April 1889 to make preparation for their journey to Gilgit.[46] In fact, Durand left Srinagar for Gilgit in the middle of June 1889,[47] a few days before the Home Government's approval to the re-establishment of the Gilgit Agency had reached Lansdowne.[48] Durand's position was quite different from that of his predecessor Biddulph, who was not only handicapped by the absence of an armed escort but also had to face much obstruction from Kashmir officials. The virtual administration of Dardistan[49] now passed into British hands through their Agent, though the Kashmir Governor at Gilgit continued to exercise civil authority over this territory.

Russian stand on Hunza claims over Raskam lands

When in the spring of 1897 the Mir of Hunza sent some of his men to resume cultivation of some tracts in the Raskam valley, this simple event snowballed into an international issue involving the three empires of the world, Britain, China and Russia. It was the action of the Chinese magistrate (*Amban*) at Yarkand in arresting the two Hunza men who had stayed at Raskam to look after their crops that actually sparked off this crisis. The matter was promptly taken up by the Mir of Hunza with the local Chinese authorities with the knowledge of full British support. The British interest in this affair was due to their suzerainty over Hunza which had been militarily enforced since 1891. They would have preferred to see this issue resolved locally through bilateral negotiations between the Hunza Chief and the Chinese authorities in Sinkiang. But seeing that the Chinese had succumbed to Russian pressure, the British extended their official support to the claims of Hunza. Russia viewed the Hunza move in exercising its rights in Raskam as part of a covert British design to extend their authority on this part of the Pamirs in their bid to outflank the Russian influence there. They saw to it that the Chinese did not give practical effect to their agreement whereby the Chief of Hunza had been allowed to cultivate this area. The two-power rivalry naturally came in handy for China which stood to gain by the postponement of the final settlement of this issue.

The rights of Hunza over Raskam and Taghdumbash are believed to have originated from the defeat suffered by the Kyrgyzs at these places at the hands of Salim Khan, the Chief of Hunza in around 1760.[50] He announced his victory and also the extension of his authority up to Dafdar through a message sent to this effect to the Chinese along with a trophy of Kyrgyz heads.[51] China expressed its happiness over the defeat of their enemies – the Kyrgyzs – by sending return gifts to Hunza. This gesture was duly acknowledged with a token present of gold dust by the Hunza chieftain.[52] Simultaneously, Hunza received the concession of cultivation, grazing and taxation rights in Raskam and Taghdumbash Pamirs from China, which, however, could not be enforced during Yakub Beg's rule. While it continued to levy taxes from the Kyrgyz settlers in Taghdumbash until early in the twentieth century, Hunza could not enforce its rights in Raskam after 1897 until 1914. The existence of Hunza forts in Azghar and the signs of cultivation at other places in this area indicated that Hunza was in actual possession of Raskam for a considerable period.

Why the Raskam question cropped up only towards the end of the nineteenth century can be explained by certain historical factors. First, the Chinese authories in Sinkiang being generally weak, could do nothing except watch Hunza exercise its jurisdiction silently over this area in the Pamirs. The local Chinese authorities did not run the risk of offending the successive chiefs of that principality who often sent foraying missions within the borders of Chinese Turkestan to enslave Kyrgyz nomads and plunder their property. Second, the Chinese derived satisfaction from the receipt of annual presents of gold dust deeming it to be a tribute from a loyal dependency, thus not bothering about Hunza rights in Raskam. On the other hand, Hunza adhered to this custom, mainly because it yielded substantial income in the shape of return Chinese presents, which were valued at about ten times more.[53] Besides, this occasion of sending emissaries to Chinese Turkestan was used to get certain outstanding matters cleared, mostly relating to Hunza rights in Raskam and Taghdumbash. Third, the need to enforce its old cultivation rights in Raskam became more acute when the people of Hunza were faced with food shortage due to the loss of income from cessation of raiding and slave trading activities. The increase in population that followed the overall peace forced the Hunza chief to search for additional avenues of income and food,[54] as there was no scope for bringing additional land under cultivation in his principality.

Coming to the point of actual Russian involvement in what was a bilateral Hunza–China question concerning small tracts, it may be pointed out that

Russia had not forgotten the extension of British authority in Hunza in 1891. The ex-Chief of this principality, Safdar Ali, while in exile in Chinese Turkestan, continued to be in touch with Petrovsky, the Russian Consul at Kashgar. Seeing that the Chinese *Taotai* at Kashgar had favourably disposed of the Hunza representation allowing the Kanjutis to cultivate some plots in the Raskam valley, Petrovsky sent a letter to the next highest authority (*Futai*), the Lieutenant Governor of Sinkiang, advising him not to allow Kanjutis to settle there as it lay on the frontier.[55] The *Futai* was also in favour of granting this right to Hunza, though he sought to allay the Russian fears by proposing to levy an annual grain tax on such cultivators so that there remained no distinction between an ordinary Chinese subject and a Hunza cultivator at Raskam.[56] So when the actual possession of seven plots at Raskam was being handed over to the people of Hunza in early 1899 by the local Chinese authorities,[57] Russia started pressurizing China against such a course of action. The Russian representatives stationed at Peking and Kashghar warned the Chinese Government, hinting about the possible Russian occupation of Taghrama (in Sarikol) as a quid pro quo to the grant of Raskam lands to Hunza. Consequently the Tsungli Yamen stalled action on the promises given to the Hunza chief and also took further steps to evict the recent Kanjuti cultivators from Raskam in a bid to save itself from the possible Russian inroads into the frontier district of Sarikol. On receipt of fresh directives from Peking, the local authorities in Sinkiang hastened to inform Mohammad Nazim Khan of Hunza about the revised Chinese decision not to allow any Kanjutis to settle in Raskam in view of the Russian objections.[58] The *Amban* (Magistrate) at Yarkand also offered to compensate Hunza for the loss of grain that would result from the abandonment of Raskam lands by Kanjutis.[59] In order to placate hurt sentiments, the *Amban* even sent two Begs from Sarikol to meet the Chief of Hunza personally and to explain the row caused between Russia and China over this issue.[60] Seeing that China had, under Russian pressure, gone back on its agreement of giving seven tracts in Raskam to Hunza, the British Government decided to resist the action. Accordingly, the British Ambassadors at Peking and St. Petersburg took the matter up with the Tsungli Yamen and the Russian Foreign Minister respectively. If Mouraviev's assurance given on 17 May 1899 to Scott, the British representative at St. Petersburg, is any indication, then Russia was not inclined to use the grant of Raskam lands as a 'pretext for acts of aggression on Kashghar'.[61] But the views of the Imperial Minister of War, General Kuropatkin, coupled with Petrovsky's pleadings from Kashghar against allowing any further extension of the British influence in Sarikol appear to have hardened the Russian stand. This is evident from their action in seeking details about the extent and position of land in question from the British Government[62] and also from the objections raised by Giers before the Tsungli Yamen in Peking.[63] Russian concern over the reported construction of a carriage road from the Indian frontier towards Sarikol was not only conveyed through the usual diplomatic channels in London,[64] but also by Kuropatkin in the course of his private conversation with a visiting British Military officer, Col. MacSwiney.[65] On both the occasions, the report was denied strongly. Kuropatkin's casual remarks made before MacSwiney on 9 June 1899 that 'if your Kanjutis go into Raskam, we shall be forced to take over Kashghar, Tashkurghan, etc.',[66] point to the seriousness attached by Russia to the Raskam affair. In such an atmosphere of mutual distrust and acrimony between the two imperial powers, China became encouraged to resist the British pressures by playing upon the Russian objections. When approached by the British representative at Peking, the Tsungli Yamen openly disclaimed the existence of any agreement between the *Taotai* of Kashghar and the Chief of Hunza over Raskam lands.[67]

While Britain and Russia were engaged in resolving the issue both in Peking and London, Scott was labouring hard in St. Petersburg to allay the Russian apprehensions. It was only after receiving a written assurance that Britain sought cultivation rather than territorial rights for Hunza in Raskam lands lying on the banks of the Yarkand river[68] (but not along the Taghdumbash Pamir as suspected by Kuropatkin), that the Russian Government agreed to drop its objections to its lease to Hunza. When Giers wrote to the Tsungli Yamen in Peking informing about his government's withdrawal of the objections raised earlier, the Yamen gave its green signal to the local authorities in Sinkiang for carrying out the lease of Raskam lands to the people of Hunza.[69] But Petrovsky was not going to let this happen. He continued to apply pressure on the *Taotai* of Kashghar by demanding the lease of Sarikol to the Russians. Even though both Giers and Mouraviev denied, before the British Ambassadors at Peking and St. Petersburg respectively, having instructed Petrovsky to stake a counterclaim for compensation in Sarikol,[70] neither did they reprimand their representative at Kashghar for re-opening the issue nor assuage China's fear of fresh Russian demands in Sarikol as hinted at by Petrovsky. Instead, Petrovsky's views were openly appreciated as 'reasonable'[71] by Giers in the course of his conversation with MacDonald at Peking. Finding another chance to defer the lease of Raskam lands to Hunza cultivators, the Chinese authorities now began to strengthen their position in and around Sarikol. They not only encouraged the settlement of Kyrgyz and other Chinese subjects in Raskam, but also terminated any remaining traces of Hunza foothold in that valley by expelling the Kanjuti cultivators. When the British Government, acting on Curzon's complaint, lodged a protest note to China on 29 May 1901 against this action, demanding the removal of Chinese settlers and also the reinstatement of Kanjutis in Raskam,[72] it received vague, unsatisfactory and evasive replies from the Tsungli Yamen. Seeing renewed British pressure being applied to China, Russia 'took steps to keep the Raskam pot mildly simmering',[73] thus forestalling any possible settlement of the issue in favour of Hunza. The Russian Ambassador in London, Baron Graevenitz, addressed a note on 14 January 1903[74] to Lansdowne protesting against the terms claimed by Satow in his notes of May and November 1901 to the Chinese. Lansdowne's assurance that the proprietary rights claimed by Hunza in Raskam and supported by Britain were not political or territorial in character,[75] appears to have answered the Russian purpose, more so because the British did not pursue the matter any further.

The roots of Russian reaction in this whole affair lay in their apprehensions about the prospective British penetration into this area in order to outflank the Russian position in the Pamirs.[76] Petrovsky must have also been worried about the growing British influence in the frontier area of Taghdumbash due to the presence of a British informer, Munshi Sher Mohammad at Tashkurghan. The Chinese recognition of Hunza claims over Raskam and Taghdumbash Pamirs was important for the British because it could be utilized to their advantage in the event of any future Russian advance towards Kashgharia. Their fears assumed serious proportions in early 1901 when Russia secured Chinese permission to station a military post at Tashkurghen postal service but actually to checkmate any future British forward move in this area. Such a clash of imperialist interests in the Pamirs from two opposite directions was behind the whole Raskam issue which remained undecided even after the Tsarist and British rules were terminated in Russia and India respectively. The issue would have been decided in Hunza's favour in 1898, but for the Russian pressure for not allowing the lease of Raskam lands to Hunza men, until in 1903 the British disclaimed any political or territorial (in other words sovereign) right for Hunza confining it to cultivation and other proprietary rights.

When in mid-1899 the Chinese authorities in Kashghar informed the Hunza chief that the decision to disallow the use of Raskam lands by Kanjutis had been arrived at in Peking after mutual discussions between the Tsungli Yamen and the British representative,[77] they only made a half-hearted attempt to solve the issue on the basis of the British offer of 14 March 1899.[78] China did not, however, make any official commitment in reply to these package proposals which involved the actual delimitation of the India–China border in this sector, thus precluding any agreed solution. After 1903, the settlement of the outstanding Raskam issue was relegated to a secondary position because the British attached primary importance to the conclusion of the Anglo-Chinese convention on Tibet and also to securing a consular status for MaCartney so as to bring him to parity with his Russian counterpart in Kashghar. In spite of their desire to sever all connections of Hunza with China,[79] the British did not abandon the Hunza claim in Raskam and Taghdumbash Pamirs. On his part the Mir of Hunza continued to levy annual grain taxes from the Kyrgyzs of Tashkurghan. It was only during the period 1899–1914 that Hunza refrained from enforcing its rights in Raskam. From 1914 onwards, when the Russian influence in Kashgharia was on the decline, Hunza adhered strictly to the practice of deputing its men to cultivate these lands, brushing aside the Chinese objections. In doing so, Hunza had received official approval from the British Indian Government. The British fully understood the strategic importance of using Hunza's rights in Raskam and Taghdumbash Pamirs to their advantage in the event of any future Russian incursions into Sinkiang. But due to their pre-occupation with other issues that were more relevant to their broader imperial interests, they refrained from applying direct pressure upon China for solving this issue. Accordingly the India–China border in Hunza and Ladakh sectors skirting Sinkiang remained undelimited even after India became independent of British rule in 1947.

Russian activity in Kashmir and its frontiers

With the British switching their active attention towards extending their control up to Hindu Kush, the Russians started evincing interest in the frontier territories of Kashmir like Gilgit, Hunza and Chitral. As already noted, Mayo's action in dissuading the Kashmir ruler from extending his authority over the frontier principalities[80] any further was reversed some six years later by Lytton in 1876 in the wake of the Russian annexation of Kokand Khanate. It was Lytton who induced Maharaja Ranbir Singh to 'obtain full control over the territories lying between them (passes south of the Hindu Kush) and the Kashmir frontier',[81] in order to secure command over such passes as were thought to be practicable for the passage of Russian forces. Thus the British gained political control over Chitral and Yasin through the Kashmir ruler and also secured the right to station an agent at Gilgit. Through the establishment of a regular agency at Gilgit under Capt. B. J. Biddulph in 1877, they monitored the Russian activities in Central Asia, Afghan Turkestan, Badakhshan and Kashmir frontiers. Such an increase in British activity near the Oxus was bound to attract Russian notice. A Russian newspaper, *Golos*, in its issue dated 3 April 1878 covered fully the observations made by Biddulph on his exploration of the Hindu Kush passes in 1876. Similarly, Kostenko in his work *Turkestan Region* published in 1880 took note of the discoveries made by Biddulph.

In 1883 news reached the British Indian Government through no other person than Aman-ul-Mulk, Mehtar of Chitral, that Russian parties then exploring Shighnan and Roshan intended to enter Chitral too. The Mehtar used this occasion as an opportunity to display his loyalty to the British by soliciting the advice of the Viceroy: 'If I should kill them or make them prisoners and send them to your Excellency.'[82] Russia was thus taking steps to keep itself informed about the developments occurring to the south of Hindu Kush. Lockhart's mission to Chitral in 1885, followed by the re-establishment of the Gilgit Agency in 1889 and Capt. Durand's activities in Gilgit were not unknown to them. Russian officer Grombchevsky's explorations in Hunza and Pamirs during 1888 and 1889 following the heels of Lockhart have already been dealt with in detail. The despatch of several British expeditions to the Pamirs and Chinese Turkestan during 1889–90 had their effect in arousing the Russian suspicions to such an extent that the Russian Consul at Kashghar, N. Petrovsky, reported to his Ministry of Foreign Affairs about the

British desire to divide the Pamirs between Afghanistan and China.[83] By now the Pamirs had landed in the centre of Anglo-Russian rivalry. The instance of the physical expulsion of Younghusband by Col. Ivanov at Bozai Gumbaz in 1891 points to the seriousness attached by Russia to the British activities in and around the Pamirs.

After the British success in occupying Hunza in 1891, events changed fast in Chitral. Aman-ul-Mulk's death in 1892 caused uncertain conditions there and the mutual strife amongst his heirs facilitated the British intervention. Nizam-ul-Mulk's installation as Mehtar with British armed assistance paved the way for stationing a British officer within Chitral territory. No sooner had Robertson and Younghusband secured a place to operate in Mastuj, than they began to advocate stricter British control over Chitral. They soon found an ardent supporter of such an active forward policy in the person of George Curzon who himself visited Chitral in 1894. The year 1895 was so eventful for Chitral that it witnessed the murder of Nizam-ul-Mulk and the consequent siege of the British force which finally culminated in the total subjugation of Chitral by the British installing Shuja-ul-Mulk as its nominal head. All these developments were fully reported in the Russian press. Perhaps under the inspiration of the Russian foreign office, *Novoe Vremya* in its issue dated 27–29 December 1892 went to the extent of claiming that Russia too had 'some interest' in the events taking place in Chitral, it being strategically important. However, the paper was fair in reporting about the Kashmir ruler's rights over Gilgit, Yasin and Chitral. Another paper, *St. Petersburg Herald*, in its issue dated 13(25) December 1892, while disputing the claims of Chinese and Afghan rights over the Alichur Pamirs, claimed Russian rights over the whole of Pamir 'up to the Hindu Kush, to the border of the independent states of Chitral, Yasin and Kanjut (Hunza)' as inherited from the Kokand Khanate's jurisdiction over these territories. The Russian official organ published from Tiflis, *Kavakaz*, in its issues dated 2(14) December and 3(15) December 1892 even anticipated a British threat to Chitral. Taking note of the British extension of the telegraph line to Gilgit and the occupation of Hunza, the paper predicted a similar fate for Chitral. Referring to the strategic importance of Gilgit and the presence of a British agent there, the paper regarded it as the 'base of a future English advance into Chitral'. Explaining the physical obstacles in the way from the Russian frontier post at Osh to Little Pamir and onwards to Chitral, the Russian columnist described the British fear of a Russian invasion of India from this side as 'absurd'. However, expressing his concern at the possible British occupation of Chitral, he wanted the Russian government to pre-empt any such move by coming to an understanding with the Afghan ruler on the issue of Chitral. According to him '[the] existence of Russian advanced posts in Shignan and on the Pamirs would place him (the Amir of Afghanistan) in a much more favourable position in his struggle for Chitral'. Another paper, *Russian News* published from Moscow, in its issue dated 8(20) February 1893 expressed concern over the consolidation of British influence in Chitral. It felt that the British were 'forestalling Russia' as England's position could be threatened only from the Hindu Kush through Chitral.

From the 1890s onwards, there was a sudden development of Russian interest in Kashmir frontiers particularly in Ladakh, Gilgit, Hunza and Chitral. The British refusal of permission to Grombchevsky for travelling down to Ladakh via Shahidullah and Karakoram in late 1889 did not daunt other Russian officers from attempting similar adventures. Since this strategic frontier territory was a sealed area even for the unauthorized British travellers, the Russians did sometimes succeed in overcoming this difficulty by securing the required permission directly from the British Foreign Office much to the discomfiture of the British Indian Government. More often, the Russians would placate London by allowing certain British officers to travel through sensitive spots in Central Asia.[84] The first Russian traveller to have succeeded in making an overland journey to Ladakh and Kashmir during this period was Prince Galitzin, who had been invited to visit Kashmir in August 1891 by the British Ambassador at St. Petersburg (R. Morimer).[85]

Finding it impossible to disown Morimer's word, the British Indian Government did allow Galitzin to enter Kashmir from Chinese Turkestan via Karakoram, Nubra valley and Leh.[86] At the same time it took all precautions to ensure that the Russian visitor would not set his inquisitive eyes on other routes towards Ladakh. In addition a British officer, Captain H. Picot, was attached with the train of the Russian Prince on his travels. After spending the winter of 1891–92 in Kashmir, Galitzin, along with Picot, left Leh on 1 June 1892 for Shahidullah (then the Chinese frontier post towards Ladakh) via the Chang La, Marsemik and Chang Chenmo passes.[87] They proceeded towards Karghalik in Chinese Turkestan via the Kilian pass. Thus the Russian visitor could have a chance to make an on-the-spot survey of two routes from Chinese Turkestan to Kashmir first via Karakoram–Nubra and Leh and then by Chang La, Marsemik, Chang Chenmo and Shahidullah. Similarly another Russian officer, Captain Novitsky of the general staff travelled through Ladakh in July 1898 on his overland journey towards Yarkand, Kashghar and Russia.[88]

The year 1899 witnessed organized trips by Russian military officers to the Kashmir frontier territories of Hunza and Gilgit. Captain Snesarev's and Colonel Polotsov's overland journey to India via Hunza and Gilgit during this year deserves special mention. While the former occupied the post of Military District Chief on the Pamirs with his headquarters at Kharog,[89] the latter had led a reconnaissance trip up the Oxus in 1894. Besides, they used the services of a Kashmiri, namely Kabir Ju (a Munshi employed at Tashkent for teaching Hindustani to Russian officers) for extracting some information from the officials employed in the British Agency at Gilgit.[90] This lends credence to Lt. Col. E. F. H. McSwiney's assertion that the mission of Polotsov was to reconnoitre the Indian north-west frontier and that of Snesarev to study the organization and administration of the British Indian Army.[91] It was not a mere coincidence that the Russian Government had permitted several British officers like Cobbold, Colonel Powell and Colonel McSwiney at about the same time to travel through Central Asia. In fact, this concession was later to be used as a quid pro quo for securing permission for more Russian officers to visit Indian borders. So when the Russian Ambassador at London sought British permission for four high-ranking military officers, Lt. Col. Shersky, Lt. Col. Kouznetsov, Count Mouraviev Amursky and Prince Curossov to travel to northern India from Ishkashim via the Dorah pass to Chitral, Killa-i-Panja to Yasin, lake Victoria to Yasin and Bozai Gumbaz to Hunza,[92] the British Government was put in a quandary. Since Staal (the Russian Ambassador in London) had cited the Russian Government's permission to British subjects to travel in Central Asia, Salisbury found it difficult to ignore his request for reciprocal treatment.[93] But the Government of India stood firm in its objection to any Russian military expert examining the passes that constituted the 'strategic defence of our northern frontier'.[94] The fact of Ishkashim, Kila-i-Panja and Bozai Gumbaz being within Afghan territory was used as an excuse by the Indian Government to avoid this complicated situation by pointing out the need to secure prior consent from the Afghan ruler.[95] However, in deference to Salisbury's wishes they consented to the Russian party travelling down from Taghdumbash to Hunza and Gilgit via the Kilik pass.[96] But as a precautionary measure, British officers were detailed to accompany the Russian party from near the Kilik pass.[97] Though this Russian mission was officially described as aimed at making ethnographic, biological and meteorological enquiries under the auspices of the Russian Geographical Society,[98] yet its composition and the routes chosen for exploration leave little doubt about the Russian desire to be thoroughly acquainted with the British frontier defence arrangements at the Hindu Kush watershed.

In April 1907 Captain Polotsov of the general staff of the Russian army appeared in Srinagar before the British Resident there asking for permission to travel back to Russia via Leh, Karakoram and Chinese Turkestan.[99] Although the Indian Intelligence Department feared that, if allowed to proceed on this journey, the Russian officer would use this opportunity for surveying the Mintaka and Kilik passes, Lord Minto as Viceroy overruled this objection, but attached a British officer to accompany Polotsov on his overland journey to Russia.[100]

No sooner had Polotsov secured the necessary permission than another Russian, M. Andriev, presented himself before the British Resident in Kashmir requesting to be allowed to accompany Polotsov to his home in Russia.[101] After his request was refused, the Russian Consul General at Bombay brought pressure upon the Indian Government to allow Andriev to travel to his home in Tashkent via the Karakoram route as a special case. Accordingly the two Russians left Srinagar for Leh on 21 June 1907, accompanied by a British officer, Major Fielding.[102] Soon after their arrival in Ladakh in mid-July, reports began to pour in about their objectionable activities. Andriev, who knew and spoke Turki and Persian fluently, was reported to have questioned traders at Leh about the extent of their trade with Tibet and Turkestan.[103] Polotsov was said to have made enquiries about the routes form Shahidullah to Mustagh and thence to Taghdumbash.[104] Fielding complained that the two Russians had surveyed the whole frontier including the Kilik pass on the pretext of going to Taghdumbash.[105] These suspicions got further strengthened by reports coming from Kashghar and Gilgit. While the British informer at Yarkand, Buniad Ali, reported that the two Russians walked in that city disguised as Andijanis so as to avoid recognition,[106] information reached Gilgit about their halt at Tashkurghan for four days to make enquiries about the Chitral and Hunza routes.[107] Subsequently, when the Military Intelligence Department of the Indian Government came to know that both Polotsov and Andriev had landed at Bombay on one and the same day and that their mission in India was controlled by the headquarters department of the general staff at St. Petersburg, the India Office was informed about the activities of these Russian officers.[108] Though Whitehall was not disposed to press the matter with the Russian Government, it cautioned the Indian Government to be more careful about granting such permissions in future.[109]

Hardly three years had elapsed when Captain Polotsov reappeared in Kashmir in April 1910, this time in the company of Prince Troubetzkoi of the Russian Civil Service, with the intention of making a return overland journey to Russia via the Kilik pass.[110] When the presence of Russian officers in Kashmir was reported by the British Resident to higher authorities at Shimla, the British Indian Government was already apprised of the matter, because the Russian Consul General in India, B. Arseniev, had made a personal call on the Foreign Department at Shimla on 26 April 1910 to seek British permission for these two officers to return to Russia by the Gilgit route crossing either the Kilik or the Mintaka pass towards Taghdumbash Pamirs.[111] Though Polotsov had met Lord Minto during his previous visit to India in 1907 for seeking similar permission which was then granted, this time he chose to avoid direct contact with the Viceroy. Instead he met the Commander-in-Chief who not only obliged with an autograph letter of recommendation, but also gave official clearance to such a visit being allowed. But Minto, whose memory had not failed him, while recalling Polotsov's 'unsatisfactory proceedings' in 1907 saw in his second visit the sole objective of obtaining military information.[112] Polotsov was refused the required permission and the Resident in Kashmir was accordingly asked not to allow the Russian 'to proceed beyond the limits of ordinary travel for visitors'.[113] The two Russians appear to have stayed in Kashmir until early September 1910 when they were reported to have left Srinagar travelling towards the Punjab.[114] Their renewed attempt to travel across Kashmir frontiers in Gilgit towards Russian Central Asia was thus thwarted.

Only a year later, the Russian Imperial Academy of Sciences started evincing interest in the collection of zoological specimens from Western Kashmir, Ladakh, Karakoram, Quetta and Sikkim. The British Government, when formally approached by Russia with a request to allow a party of

Russian entomologists to make such a journey in India,[115] was put in a fix. While the India Office, being averse to allow the Russians any access to frontier regions in Sikkim, Ladakh, Karakoram and Quetta, proposed to collect the required zoological specimens through the agency of the Indian Government, the Foreign Office found it difficult to refuse such permission.[116] However, the matter was resolved when the Indian Government gave its clearance to the Russian expedition, if it travelled via Rawalpindi–Murree–Baramulla–Srinagar–Sindh–Zojila–Dras–Kargil–Leh–Karakoram towards Kashghar.[117] The Russian party comprising Avinov (gentleman-in-waiting to the Tsar), Mamaev (Attaché to the Russian Ministry) and Jackolson (Member of the Imperial Russian Geographical Society) spent the summer of 1912 in Kashmir and Ladakh from where they travelled back to Russia via the Karakoram pass, Shahidulla and Kashghar.[118] The activities of this Russian expedition appear to have been restricted to that of catching butterflies in the upper reaches of Kashmir and Ladakh, on account of the 'unobtrusive watch' being kept by the Resident in Kashmir over the party. But the inclusion of Quetta in the first Russian programme could serve no better purpose than assessing the British strength at the Quetta, Rawalpindi and Attock military stations.

Following the increased British activity in Gilgit, Hunza, Nagar, Chitral and the North Western Frontier Province, the Russians made sustained efforts to make an on-the-spot survey of the British military arrangements by despatching experienced officers of the Russian General Staff to that area. In 1897–98, one such officer, Novitsky, not only succeeded in examining the forts lying on this frontier, but also in returning to Russia overland by the Leh–Yarkand–Kashghar route. In his semi-classified publication *Military Sketches of India* brought out in 1899 (in Russian), he opined that this frontier belt was not fortified in a way to meet the much publicized Russian threat to India.[119] Taking due note of the weakness of most forts to withstand the artillery attack of a modern European army,[120] Novitsky concluded that the British cry of a Russian menace to India was meant to be used as a pretext for its own annexationist policy.[121] To support his point, he drew a comparison between the British expansion towards India's north-western frontier and that in Burma, eastern China, Sudan and Fashoda.[122] Similar conclusions had also been drawn by Lt. Col. Kornilov after he visited northern India in the early twentieth century. According to him, Rawalpindi was the only well-guarded place in this area.[123] It was Col. Polotsov who, after his first visit to India during 1899, gave vent to his views in a book *North Western Frontier of India* (in Russian). He wrote in favour of the Russian army crossing over Afghanistan and the Indus river towards Punjab and Delhi, if war broke out between Britain and Russia elsewhere, a task which he believed was realizable under certain international conditions.[124] Soviet scholar N. A. Khalfin has quoted from a confidential note dated 8 December 1902, prepared at the Russian military headquarters, to say that Polotsov's proposals were dismissed as adventurist, undesirable and impracticable by the highest military authorities including General Kuropatkin.[125] Polotsov's opinions were of course based on his individual assessment of the British military strength, as he had not only conducted a survey of the Peshawar and Quetta district,[126] but also travelled via Gilgit during 1899. By repeating his visit to India in 1907 and again in 1910, with the sole purpose of making an overland journey to Russia via Ladakh, Karakoram, Gilgit and Taghdumbash Pamirs, Polotsov obviously did so at the behest of the Russian General Staff. During his stay in India, he used to report about the general political developments to the Russian Foreign Office.[127] So the British suspicions about his working under the directions of the Russian Government, though too late to prevent his journey beyond Ladakh in 1907, were not unfounded. Accordingly in 1910, he was not allowed to go beyond Kashmir.

The routes and areas chosen by the Russian officers, whether Novitsky, Kornilov, Polotsov, Avinov or others for their field trips in British India, present an invariable similarity. Their preference to traverse the frontier areas in and around Ishkashim, Chitral, Yasin, Hunza, Gilgit, Quetta, Peshawar and Rawalpindi was guided by the intelligence requirements of Russia.

Such reconnaissance missions were not a prelude to any Russian imperialist designs over northern India. It becomes clear on the perusal of Russian official opinions on such an adventure as are cited by Khalfin. In the eyes of the then Russian Foreign Minister, V. Lamzdorf, India was important for Russia as it represented a sensitive nerve of Great Britain, which when touched could force the British to drop its hostile stance.[128] Similarly General Kuropatkin, in contrast to the general British opinion of being an ardent supporter of Russian advance to India, is reported to have advocated alliance with Britain in order to achieve a peaceful agreement for partition of Asia between the two powers.[129] In fact, during 1912 Russia made a categorical declaration to the British Government to the effect that it had neither any desire to take over the administration of Chinese Turkestan, nor any wish to have any footing in Kashghar or the neighbourhood of the Kashmir frontier.[130] What then had necessitated the despatch of such experienced military officers to British Indian frontiers? It appears that Tsarist Russia had not closed its option to pressurize Britain by a mere show of force at India's north-western frontier, as and when such a necessity arose in order to relieve itself in Europe. And for judging the expediency of such a move they had but to be equipped with the up-to-date intelligence about British military strength in this area, more particularly after a series of British frontier campaigns in Hunza, Nagar, Chitral etc. In this context it is important to note that the Russian Government had even started a special two-year course in the Hindustani language for its military officers, who were required to acquire field experience in India later.[131] The whole matter of Russian intelligence activities in India needs to be viewed in the context of the Anglo-Russian rivalry both in Asia and Europe. By stationing a military commandant with an armed escort at Tashkurghan, Russia only sought to counteract the influence created there by the native British informer. If this officer, in conjunction with the Russian Consul at Kashghar, used to employ native agents to obtain intelligence from Hunza, Gilgit and Chitral territories,[132] the practice was not unknown to the British who even used the services of the Agha Khan to extend their influence over the Ismaili residents in Wakhan and Sarikol. That the Russian officers commanding the Pamirs also indulged in such intelligence activities was admitted by one such officer, Lt. Kevekiss, before a visiting British officer, Ralph P. Cobbold, who travelled through Gilgit, Hunza, Pamirs and Chinese Turkestan in 1897.[133] According to Kevekiss, several Tajiks were employed by the Chief Political Officer at Kharogh for collecting intelligence from Chitral and Gilgit.[134] Even the Prime Minister of the Mehtar of Yasin was reported to be supplying the Russians with information.[135] In this 'Great Game' both powers were playing their role to the best of their ability employing all means within their power short of armed conflict. If the Russian officers got lured to travel overland from India to their country across the high snowy peaks of the Karakoram ranges in Ladakh and Gilgit, the British too were eager to grab every opportunity to travel within Russian Central Asia. Both powers sought to outwit each other by conceding stray permissions to such visits by their officers, just in the hope of receiving reciprocal treatment. So the Russian consul at Calcutta, L. K. Reweliotty, was able to make an overland journey to his home via Gilgit, the Kilik pass and Kashghar during the summer of 1913, only after two British officers had secured permission to travel in Russian Turkestan.[136] On one occasion, when the need to obtain military intelligence about the Russian strength at Kushak was felt, the British were prepared to allow the Russian military officers including those commanding the frontier pickets in Pamirs, to make an overland trip to India via Gilgit and even permit them to see Malakand and Khyber.[137]

The Anglo-Russian convention of 1907 which brought a relaxed atmosphere, could not totally eliminate the deep-rooted mistrust between the two erstwhile rivals. However, World War I brought the two countries together to meet the common threat. As a result, even the imperial agents posted in remote pickets like Tashkurghan and Kashghar (both in Chinese Turkestan) used to work in unison on several matters of common interest, quite contrary to their constant mutual strifes. But this brief period of mutual cooperation came to an end on the eve of the October Revolution in Russia. Now the British geared their imperial machinery not only to prevent the entry of Bolshevism into India but also to destablize the Soviet power in Russian Central Asia.

3 The Gilgit dimension of the Kashmir frontier

P. N. Jalali

There is a Gilgit dimension to the Kashmir issue, a dimension that brings to mind a hundred-year-long unending debate which highlights the strategic importance of the Himalayan ramparts, in the Pamirs–Karakoram–Hindu Kush trijunction, where the huge land masses of the five nations meet. This huge land area is of great historical import and is known by yet another name – the Northern Areas, a place where many pathways of culture and civilization have criss-crossed, an entrepot of many cultures with rare complexities and sophistication. It is here that Gilgit manuscripts, unravelling among other things mass killings of Buddhists, were also found. In a way, it symbolized the end of a long search for the scientific origins of the British Indian empire along the Indus in the north; the triumph of colonialism over orientalism – a marriage of British colonialism and feudal despotism. It is an area with a variety of ethnic characteristics, such as 'Kafirs' of 'Kafirstan', a tribal sect with weird ritual dances and animal sacrifices. Kafirstan, in addition to other tribal entities living in these Himalayan ramparts, poses a challenge not only to Islamic nomenclatures, which are sought to be imposed under the new Islamic dispensation, but also symbolizes an anguished cry against suppression of tribal autonomy in the name of radical Islam groaning under military rule.

Closed to the prying eyes of the world, the so-called Northern Areas, constituting the last remnants of colonialism in South Asia, off and on burst into the open with a massive outcry against the stranglehold imposed on its inhabitants by the military rulers of Pakistan. Perhaps few people outside this subcontinent know that what is now known as the Northern Areas of the erstwhile state of Jammu and Kashmir – comprising Hunza, Nagar, Punial, Yasin, Kuh, Ghizar, Ishkoman and Chilas – is the single largest territorial division of the state. Along with Ladakh, the frontier area running parallel to the Karakoram ranges from the Hindu Kush to Changthang in the east, constituting 63,544 sq. miles (164,500 km^2) out of a total area of 84,471 sq. miles (218,780 km^2) of the entire Jammu and Kashmir state.

With a population of nearly two million, the Gilgit region has been in focus as an important milepost on the famous Silk Route in the pre-colonial period. What was mainly a route for a brisk exchange of trade and cultures and conquests by Mughals and predatory raids by Ghaznavis, beginning with Alexander's campaign, underwent fundamental changes. The Silk Route became a road of colonial confrontation between Tsarist Russia expanding to the south and the British colonial power rushing to the Indus line in search of a geographic border for the empire. Gilgit and its environs, especially the principalities of Hunza and Nagar, became milestones of the limits as well as the expansion of the British colonial rule. The region witnessed the rise and fall of great civilizations, such as Khwarezm in the heart of Central Asia, the Kushan empire, and the intermingling of many great religions such as Hinduism, Buddhism and Islam.

The British colonial focus on the Gilgit region as a vital strategic outpost of the empire began early in 1817 when expanding British power reached the Sutlej with Ranjit Singh standing firm to stem their advance and Sindh Amirs holding forth in the south. This provided the key to the Indus navigation, which was the lifeline for a huge amount of trade passing from the Central Asian Khanates over the northern ramparts and down to the Kashmir valley, through Ranjit Singh's Punjab.

The British colonial drive towards the north set a process in motion to find the navigability of the Indus. The 1809 Treaty, which brought British power to the north and the north-west Indus, made navigation possible with boats, as well as handling goods from Central Asia between Kandahar and Kabul in order to escape the high taxes imposed by Ranjit Singh; in fact, to control exports through sea routes to European markets.

Occupation of Sindh and Ferozepur, therefore, became the first priority for the advancing British, who were highly nervous of the outcome of the Turkmanchi Treaty between Russia and Persia. The British thought that the Treaty had threatened the internal security of the Indian empire. Following this Treaty, not only in Khorasan, Afghanistan and Turkestan, but also in the bazaars of Bombay, was the advance of the expeditionary armies of Russia and France openly discussed, the armies first pushing to Khorasan and then to India.

Ranjit Singh, who was trying to utilize the intercolonial conflict between Pax Brittanica on the one hand and the Franco-Russian entente on the other in his favour, was trying to take over the possessions of the Sindh Amirs and the right to acquire a major portion of Afghan possessions in India, especially the Kashmir valley, along the Silk Route right up to Gilgit. What unnerved the British was Ranjit Singh's bid to expand his kingdom to Sindh and seize the entire trade corridor to the Arabian Sea, from where Sikh rulers would establish contact with the two foreign powers, Russia and France.

The Sindh amirs, who were conscious of the covetous eyes of the British on their possessions and the Indus, were constantly beseeching Ranjit Singh to unite to stem the British tide. It was at this critical juncture that the British Secret Committee of the Honourable High Court of Directors met and stressed the need for obtaining free navigation of the Indus with a view to securing the advantage that would result from substituting its influence for that of Russia, through its commercial intercourse with Bukhara and the regions lying between Hindustan and the Caspian Sea.

Aware of the high stakes involved in seizing control over the Indus for navigation, the British persuaded two Sindh amirs, Rustam Khan, the Amir of Khanpur, and Murad Ali of Hyderabad, to enter into a one-sided treaty and as proxies to throw the Indus open to the British military and political penetration. Under the treaty, signed reluctantly by the Sindh amirs, the British obtained the right to take the final decision regarding taxation duties on goods passing through the Indus to the Arabian Sea. This was a variation of gun-boat diplomacy with which the advancing British seized the veto in their hands to control the entire trade flow from Central Asia to Sindh and thereby the right to manipulate the economies of the regions and the states, including that of Ranjit Singh whose removal was now on their agenda.

The next step was to coerce Ranjit Singh to sign the treaty, opening Sutlej to commerce as well. The British agent, Captain Wade, told the Maharaja that the expansion of trade in his (the Maharaja's) dominions would result only after signing the treaty. But the British agencies had disrupted the economy of the Sikh state by imposing a heavy toll of Rs. 570 on the boats carrying large loads. This included the pashmina shawl output from 20,000 looms that were operating in Kashmir. The shawls were exported mainly to the European market, especially France, where they posed stiff competition to woollen manufactures from Britain.

Ranjit Singh signed the treaty on 26 December 1832. With that he signed what was virtually a death warrant for the Sikh state which stood between British ambition and drive to reach the Indus line on the one hand and Indian resistance against foreign domination on the other. Just 15 years after signing the treaty, Ranjit Singh and his state, which presented a formidable challenge to British power, had vanished from the scene, yielding place to the creation of a new independent state of Jammu and Kashmir under Maharaja Gulab Singh. The new state was created to fulfil the historic task of taking the triumphant British flag to the northernmost corner of the subcontinent without any obligation on the part of the British to confront Tsarist Russia directly, which in turn was seeking to find an access to sea trade via Persia (Iran) and Afghanistan.

Even though vanquished, one cannot but give credit to Ranjit Singh for having understood the historic importance of colonial conflict between the British power on the one hand and Franco-Tsarist power on the other, and accordingly planning his moves to build a regional power base, which failed to materialize because of the superior diplomatic, economic and military organization of the British. Ranjit Singh was the only ruler, after Tipoo Sultan, who understood the superiority of the British and tried to establish his army on the Western model, utilizing French military talent to remodel his artillery. He introduced the European style of warfare. It was too ambitious a task, though, and before his efforts could yield tangible results, the

British had succeeded in subverting his regime, which lay dismembered immediately after his death.

The British persuaded the Sikh monarch to expand towards the north to Kashmir and Tibet, rather than to go south for an outlet to the sea. Having been successfully barred from taking over Sindh, the Maharaja took over Jammu and Kashmir including Gilgit, with the connivance of the British. The popular belief that Ranjit Singh was prompted by Raja Kak Dhar's pleadings to deliver Kashmiris from Afghan despots might have also influenced the decision to take over Kashmir. But the fact remains, it was the British who successfully persuaded Ranjit Singh to wrest Kashmir from the Afghans; first to contain them within the Indus line in their own territorial environs and secondly to put an end to the repeated Afghan forays and armed campaigns. Ranjit Singh's reign, when the British colonial rule was in the process of consolidating the empire and as a consequence forging a nationhood (in the modern sense of nation-state) of Indians from Brahmaputra to the Indus and from the Arabian Sea to the Himalayas, needs to be reappraised in a larger context.

Dismemberment of Ranjit Singh's state and replacing it with the Jammu and Kashmir state, given to Raja Gulab Singh as an 'independent possession', was qualitatively a new experiment in forming a vassal state in a strategically important region of emerging India. The British had hitherto been suppressing the traditional structure of the states in India to consolidate their own power. The creation of the Jammu and Kashmir state was primarily meant to fulfil the historic task of consolidating new imperial power along the Indus and Karakoram to demonstrate the arrival of the empire to the nations beyond. It gave one crucial advantage to the British. They did not have to show their own hand and act in their own name on the crucial border, but through a proxy – who clearly understood that the survival of the new state was closely linked with their readiness to mobilize all their necessary resources to expand northwards and build up a security network in the region.

The scions of the Dogra dynasty, beginning with Raja Gulab Singh and ending with Maharaja Hari Singh, fulfilled their assignments without any reservations, but the strategic requirements of the empire and its global commitments did not always coincide with their own interests and motivation, resulting in conflicts. For instance, during the early years of the twentieth century, Maharaja Pratap Singh was deposed to make room for a direct takeover by the British under the garb of a council of officials headed by Raja Ram Singh. This sordid intrigue was exposed by Amrit Bazar Patrika as an unjust step by the British to deprive an Indian prince of the right to rule his state without their interference. But behind this unprecedented step were the lingering British doubts about the role of the Dogra Raj as dependable guardians of the strategic frontier in the face of advancing Tsarist Russia.

In a heated debate lasting over a decade, the British strategists had been assessing whether a direct takeover in Kashmir would not serve the supreme interests of the empire, rather than leave it to the care of a Raja, who might prove a serious impediment in times of a crisis. One section continued to adhere to the view that Dogra administration with its misrule was the best bet in the sensitive region for the British, while another section with powerful backers in the Governor General's office felt that the time had come to dispense with the Maharaja's rule to pave the way for direct colonization of the Kashmir valley, with its Mediterranean climate, for settling the families of British troops permanently who had otherwise to go to England for holidays. A permanent abode for the British in Kashmir and their troops would also meet the imperatives of the defence of the empire threatened by Tsarist Russia.

A simulated campaign was launched to the effect that eastern possessions would be lost if early preventive measures were not taken to stem the Russian advance. The military assessment of the new threats focused on the vulnerability of the two regions, one the Afghan border from the Wakhan along the Wakhan tongue and the other through the Dardanelles to the Mediterranean with the Russian navy making a thrust via the Dardanelles to challenge the superiority of British navy in the Mediterranean Sea. If the Russian navy confronted its British counterpart and by chance gained an upper hand, and simultaneously Russian troops crossed the Afghan border, the event would signal a general revolt in the east to defeat the empire.

The only counter-action against a Russian advance, the British strategists surmised, was to descend from the Wakhan heights through the Afghan corridor, simultaneously cutting off the Russian navy's thrust into the Mediterranean and thereby stemming the tide of an Indian revolt. This exaggerated view of the Tsarist threat could be met effectively by the direct takeover of Kashmir and by stationing British troops and their families permanently in the valley. An immediate offshoot of this understanding was the devising of plans for linking Kashmir by railway with the plains area, mineral surveys for the possibility of industrial development and strategic road connections, and in particular construction of a road to Gilgit. On the cultural and ethnic sides, researchers began studies of the Jewish origin of the Kashmiris and their belief that Jesus had chosen a burial place in Kashmir.

Hence the overthrow of Pratap Singh to pave the way for the implementation of grandiose colonization plans. But soon afterwards better sense prevailed, forcing a change of priorities, and the old order was restored, but with a firmer grip by the Resident. This manifested itself in the Governor General controlling the trade flow from Central Asia into the state, and acquiring the right to cripple the fragile but lucrative woollen and other manufacturing trades from Kashmir and from both Chinese and Russian parts of Central Asia.

From the beginning of the twentieth century to the end of the British Raj, the strategic focus lay on Kashmir and Gilgit and revolved around the assessment made in the early part of the previous century with varied emphasis to suit the exigencies of the situation as it developed in the region. This assessment largely remained unchanged between 1846 and 1935, divided into three distinct periods. The first, up until 1870, was the formative period. Then between 1870 and 1885 was the expansionist period, when Maharaja Ranbir Singh took the state flag to Chitral, Hunza and Nagar, the last two outposts opening into Chinese Turkestan. The final phase, which witnessed drastic changes in the relationship between the Dogra rulers and the Governor General, lasted from 1885 to 1935, when the British took over the Gilgit Agency in the wake of a popular anti-autocratic upsurge in Jammu and Kashmir.

With the emergence of a strong valley-based popular movement against autocratic rule, a new factor in the situation was craftily utilized by the British to seize important concessions, such as a lease of the Gilgit Agency. Initially they identified themselves with the popular aspirations against the autocratic misrule perpetrated by the burdens imposed on the feudal administrative structure by the requirements of imperial expansion towards the north. The military expeditions carried out to consolidate power on the northern frontier, Gilgit, Hunza and Nagar, had a direct bearing on the agriculture in the state. The expenditure on such expeditions reached 24 per cent of the total revenue of the state, whose mainstay was the forced labour employed to cultivate land through the system of *Khudkasht* – a system of running state farms as a part of the village land itself. The forced labour by the cultivators remained unpaid.

While in the Kashmir valley massive *begar* (slave-like forced labour) was introduced to carry loads for expeditionary forces on the inhospitable Gilgit roads, in the Dogra land of Jammu, the *leri* system was introduced to pay sepoys by remission of rent and not through regular salary. Each group of ten houses in Jammu was obliged to supply one sepoy under forcible enlistment for serving with troops without salary on Gilgit expeditions. The families of those who deserted and failed to enlist were forced to provide substitutes and in case of failure, severely punished. These oppressive measures devastated the agriculture and unleashed unrest, which was suppressed through a network of revenue menials and officers, including women, who were used as spies to keep a vigilant eye on agitators. But when ultimately the popular anger burst, the British utilized it to pressurize the Maharaja to seize strategic concessions and to appear as promoters of justice and fair play, projecting the struggle against autocracy and colonial rulers as a struggle of oppressed Muslims against non-Muslims in the state. Though the British failed to influence the course of freedom movement in the state, it would be an understatement to say that their ideological influence on it had been eliminated.

The British emphasis on this strategic aspect of Kashmir and Gilgit on the eve of India's freedom was highly disproportionate to the overall urge for peace and stability among the people in the region. The direction that the British policy would take was evident during the visit of the Cabinet Mission to Kashmir, when it declined to discuss the issue of freedom of over eighty million people of princely states as raised in a comprehensive memorandum presented to it by the National Conference (NC). The NC demanded that when power was transferred finally from British hands to the people of India in the princely states, the sovereignty should be transferred to the people rather than the princes. Far from discussing this burning issue with the accredited representatives of the popular movement in the state, the Cabinet Mission sent military experts to Gilgit to reassess its strategic importance. Their visit to the Northern Areas was a confirmation of the fact that the British policy towards the freedom of Jammu and Kashmir would be mainly influenced by their judgement as to which party, in their opinion, would carry out the strategic task of keeping Jammu and Kashmir in the Western global network as part of the reorganization of the Indian subcontinent into two separate dominions and if possible a buffer of princely states. It could be used to influence the destiny of both the dominions, especially the Indian Union, which did not endorse the Western strategic perceptions as befitting a free country of India's size.

The Maharaja of Kashmir, principal actor in the process of transition to freedom in Jammu and Kashmir, began to work on the British Resident's perception that the state could not remain independent or alternatively join Pakistan. Prime Minister R. C. Kak, Army Chief Brigadier Scott and Police Chief Powell did influence Maharaja Hari Singh for not taking a decision on the accession issue well in time before the partition. While Brigadier Scott disbursed the state army in pickets all over the border, Kak kept on feeding the Maharaja with lengthy notes providing 'solutions' to problems which might arise in the event of a state not acceding to either of the dominions. The notes included ideas on arrangements for procuring military and defence equipment, state operation of its own airlines, and how to conduct external relations with the two dominions. While all the openings with New Delhi were completely concluded, those with Pakistan were kept open, allowing armed bands to conduct probing forays into the state territory, beginning on 5 September 1947.

As a prelude to preparation for sending hordes of armed tribesmen to seize Kashmir by force, the British Governor of the North West Frontier Province (NWFP), Cunningham, was personally directing the campaign to mobilize armed levies, while senior British generals kept the secret close to their chest, until they were forced to admit that they had the prior knowledge of a Pakistan-backed attack on Kashmir, which was publicly announced on 22 October 1947 with the capture of Muzaffarabad on the Srinagar–Rawalpindi road. This treacherous attack on the state took place despite repeated pleas by the Maharaja's Government to the Pakistani Government to halt the marauding tribes so that the state could decide on its future affiliation in calmness and tranquillity. Parallel to this was the stand taken by National Conference leaders who asserted that a decision on accession could be taken only after people were free from the stranglehold of autocracy. Released from jail after a long civil disobedience movement, Sheikh Abdullah appealed to Pakistani leaders to respect the resolve of Kashmiris to attain their freedom first and then afterwards to resolve the issue of accession. The Pakistani leaders responded with a massive armed intervention, which triggered off a popular resistance upsurge never seen before in the state.

Joined by the Indian army a few days after the Maharaja acceded to India, the popular resistance, comprising thousands of artisans, factory workers, students and members of the National Conference along with the National Militia, a voluntary organization, spearheaded the popular campaign against Pakistan. Finally, the tide was turned as the Indian army in swift actions forced the raiders to flee the Kashmir valley, but not without committing unheard of atrocities on the innocent people, indulging in looting and burning apart from destroying property extensively. Pakistani rulers described these freebooters as freedom fighters, but the myth soon exploded, as their armies joined the fray to tilt the balance in favour of the retreating tribal raiders.

While in the Kashmir valley and the Jammu region Indian and Pakistani armies were confronting each other on a wide front, far-reaching changes were taking place in the strategic Northern Areas, especially Gilgit, which had been restored to the state after being administered directly by the Governor General of India since 1935. Just prior to its restoration, Hunza and Nagar, the two principalities, had renewed their pledge of loyalty to the Maharaja and the state of Jammu and Kashmir. The Mirs of the two principalities, which had figured in a prolonged dispute, due to a suggestion that the two formed independent states outside the administrative jurisdiction of the Maharaja, came to Srinagar in July 1947 exchanging gifts and traditional courtesies, symbolizing that their territories were part of Jammu and Kashmir state.

The day restoration took place, Mahatma Gandhi was visiting Srinagar in a gesture of solidarity with the fighting people of Kashmir and their leader Sheikh Mohammad Abdullah, who was in jail. On being told about the reason behind the jubilation witnessed in the streets on 2 August 1947 on the restoration of Gilgit, Gandhi commented that it would have been better if Gilgit had been awarded local area autonomy to govern itself and to preserve their traditional ways. But this was not to be. Pakistani rulers with the active connivance of the British had different plans for the region.

Prior to the outbreak of a local rebellion led by Major Brown of the Gilgit Scouts, the state Government maintained a garrison of one battalion of infantry and one mountain battery, besides the Gilgit Scouts maintained by the Political Department of the Government of India. Immediately before the partition, fearing trouble, the detachment at Bunji was increased by a further two platoons. The placement of these forces in the sensitive region was considered sufficient for maintaining internal order, which was planned to be reinforced by deputing Brigadier Ghansara Singh as the new Governor, after restoration of the area on 1 August 1947 to the state Government.

The Prime Minister, R. C. Kak, far from acting quickly to depute the newly appointed Governor to take over the key assignment, delayed his departure by one month, during which time the local rebellion had succeeded, and when Ghansara Singh attempted to take over, he found himself a prisoner at the mercy of rebels. Even so the state Government and the Government of India, after signing the Instrument of Accession, treated the rebellion and issues arising from it as a local matter to be resolved mutually. Pakistani troops were nowhere on the scene until then, nor were any personnel of the administration of Pakistani dominion present to exploit the turmoil and annex the territory, which they did later.

Unlike other areas of the state, namely Mirpur, Mendhar, Kotli Bagh, Poonch and Muzaffarabad, where Pakistani troops acted in concert with tribal raiders, the Northern Areas were declared a free territory by the rebels who later handed over the area to invading Pakistani troops under a sordid deal which had no sanction under either the Indian Independence Act or international law. It is of interest to note that until the Karachi Military Conference of July 1948, Pakistani rulers firmly adhered to the stand that their troops were nowhere on the scene in Jammu and Kashmir. Those fighting inside the state, they maintained, were 'freedom fighters' in whose name they also declared the formation of an 'Azad Kashmir' Government based in Muzaffarabad.

Claiming to have 'liberated' the territories from the Maharaja's administrative jurisdiction, the personnel of the so-called Azad Government should have asserted its administrative jurisdiction and as a sovereign entity responded positively to repeated appeals from the Indian state of Jammu and Kashmir for the peaceful settlement and solution of issues involved. In fact, the emergency administration under Sheikh Abdullah deputed emissaries like Captain Akram with an offer of talks, but the Azad Government acted as perfect mercenaries for Pakistani rulers, who at the Karachi Military Conference now accepted the presence of Pakistani troops fighting in the state against Indian security personnel.

Pakistan's admission about the presence of its troops introduced a material change in the situation, for without this admission their bid to enter themselves as a party to the Kashmir dispute would have misfired and Western interventionist forces lost a mascot in whose name they meddled throughout the last 60 years. The Karachi Truce Agreement, under which a ceasefire was ordered, noted that Pakistan had agreed to withdraw its troops from the state. The Government of Pakistan, the Truce Agreement further said:

> will use its best endeavours to secure the withdrawal from the state of Jammu and Kashmir of tribesmen and Pakistani nationals not normally resident therein who have entered the state for the purposes of fighting, that the territory evacuated by the Pakistani troops will be administered by the local authorities under the surveillance of the UN Commission for India and Pakistan (UNCIP).

The Truce Agreement ended with the assurance that once Pakistani troops were withdrawn, the Government of India would maintain its troops necessary for law and order. The authority of the Jammu and Kashmir Government for safeguarding the law, order and peace was also upheld.

As regards the future of the 'sparsely' populated and mountainous regions of the state in the north, the Prime Minister of India in a letter to the UNCIP which finalized the Karachi Truce Agreement, asserted that the authority of Jammu and Kashmir 'has not been challenged or disputed, except by roving hostile bands of hostiles, and once the Truce Agreement is implemented we must be free to maintain garrisons at selected points in this area for the dual purpose of preventing the intrusion of tribesmen, who obey no authority, and to guard the trade routes from the state into Central Asia'.[1] Responding to the Prime Minister's letter, the Chairman of the UNCIP, Joseph Korbel, in his letter of 20 August 1948 to Jawaharlal Nehru, assured him that the matter of defence administration of the Northern Areas would be considered in the implementation of the 13 August 1948 resolution.[2] But the Commission later resorted to subterfuge declaring that it would study the situation in those areas. Bringing this fact to the notice of the Security Council, one of the Commission members, Dr. Chyle, submitted a minority report on 1 December 1948, giving some interesting insights on the inner workings of the UNCIP, which was, in fact, heavily influenced in its decisions and judgements by Western strategic considerations and designs. Dr. Chyle commented upon 'the obligation of the Commission which due to the Commission's own faulty judgement did not represent any formal part of the resolution of 13 August 1948. The Commission now explains as a declaration of intent to study later on the situation in the north.' But by the time the Commission decided to study the situation, Pakistan, contrary to the provisions of the 13 August 1948 resolution, had conquered many places in northern Kashmir to present a fait accompli to the Commission. Dr. Chyle in his report to the Security Council explained that 'the situation in the Northern Areas had meanwhile undergone material change in that the Pakistani Army had conquered many strategically important places during the interval. The Commission is bound to admit that while the reservation of the Government of India of 20 August 1948 may be legally valid, it is physically impossible to implement it.'

Since then, the occupation of the Northern Areas by Pakistani troops has ended in annexation of the strategic area, with Pakistani military rulers denying it was a part of Jammu and Kashmir state, not even of the Islamabad controlled so-called Azad Kashmir Government. In the lexicon of the Pakistani ruling establishment what exists is only a 'Northern Area', not ethnic entities like Gilgitis, Hunzaits, Baltis and other virile tribes and people whose history is older than Pakistan itself, people who cry out loud for the recognition of their nationhood and their right to self-determination.

4 India's Himalayan frontier

Strategic challenges and opportunities in the twenty-first century

Vijay Kapur

Introduction

The discipline of political science conceives of nationhood in terms of tangible attributes that denote territorial control and a definite and recognizable political identity. Primarily among these is the notion of a clearly delineated frontier that provides a sense of security to the inhabitants and also has the capacity and wherewithal to keep unwelcome intruders at bay. Once again, the differentiation between a natural and an artificial frontier is significant, with the geography associated with a natural frontier considered more defensible and psychologically impregnable than a purely man-made frontier. The contrast between the pre-World War II Maginot Line that divided France and Germany and the English Channel is of significance in this framework. The German army (Wehrmacht), which breached the Maginot Line in 1940, utilized the 'blitzkrieg' approach of swift movement, overwhelming force and outflanking manoeuvres to render the Maginot Line helpless in stopping their swift movement into France. On the other hand, the English Channel has served as a natural barrier between England and its adversaries, last being breached by William of Normandy in 1066. Similarly, the United States of America has benefited strategically from the fact that the Atlantic and Pacific Oceans lie on its eastern and western flanks as natural frontiers. It is significant that the war of 1812 between England and the United States of America was the last occasion that a foreign army had fought a war on the American mainland. Successive American administrations postulated the idea of strategic security for the American nation state thereafter through a dynamic but consistent policy which fostered the development and maintenance of strong neighbourly relationships with the nations of Canada and Mexico located on their northern and southern flanks, while also maintaining naval dominance as a form of deterrence on their eastern and western seaboards. It is in this contextual framework that the Himalayan frontier of India is analysed and discussed here.

The Himalayas have provided a natural frontier for the Indian nation since antiquity. But a lack of clarity about methods enabling the development of an integrated Indian national identity as well as a misplaced appreciation of strategic realities led to its progressive erosion as a viable frontier. Paradoxically, it was the onset of British colonial rule, with all its manifestations in the post-1757 period, which saw the emergence of renewed attention on reviving the natural security qualities of India's Himalayan frontier. The goal of this exercise was more to protect British colonial dominance within the Indian subcontinent than to accomplish the security of the citizens of colonial India. This approach, which suggested threats to the British colonial edifice in India not from within the local population but from outside its borders, reached its logical conclusion during the Viceroyalty of Lord Curzon (1899–1905). The Curzonian paradigm of governance was aggressively despotic, but instilled both benevolent and non-benevolent components. This sometimes pitted him against the British Government in London when he felt that colonial India's interests were not being safeguarded. The fear of expanded Russian influence in India's strategic neighbourhood was always a prime motivator for Curzon's policies and frequently led him into conflict with his counterparts in Whitehall. His efforts to continually rebut the assumption behind British policy emanating from Whitehall that India's subordinate status required only acquiescence, not consultation and respect, formed a part of this continuum. His resistance to efforts by the Balfour Government to make the Indian colonial administration pay the cost for part of the British army contingent located in South Africa, but notionally earmarked for stopping a future Russian entry into the Himalayan region adjacent to India, is an interesting case study in bureaucratic jousting. His comments in a letter dated 30 July 1903 to the then Secretary of State for India (John Brodrick) state:

> I cannot conceive anything more unfortunate than the impression which British Governments are steadily building up in India, that India is always to be treated from the selfish and Shylock point of view by the people at home and that it is to the Viceroy alone that she can look for defence of her interests. This puts the Viceroy into a position of quasi antagonism to the Government at home, of which you are always telling me that he is a colleague.

Curzon was convinced that protection of the British colonial state in India required dominance of the strategic Himalayan neighbourhood to stem any incursion by other interested regional state actors like Russia and China. The idea of developing 'buffer states' formed an integral part of this strategic philosophy and the crux of efforts to develop an integrated external policy paradigm during his term in India.

The Curzonian paradigm – buffer states and military-driven outreach

Curzon's much talked about conflict with Kitchener from 1902 onwards was as much a product of the proverbial struggle for dominance in policy-making between the civilian and the military branches as a reflection of contemporary contradictions in colonial foreign policy objectives. Kitchener's observation, in a communication of December 1902 to the then Commander-in-Chief of the British army, Lord Roberts, that the Indian army was unprepared for a large-scale conflict and that major operations and positional changes were needed to make it capable of stemming any perceived Russian threat, only served to widen the steadily increasing gulf between Curzon and Kitchener. Obviously Kitchener felt that he and Curzon were on an equal administrative footing and policy-making within the colonial setting of India was accomplished through a 'Concordat' framework between the military and civilian branches. On the other hand, Curzon was firm in his belief that the military operated under his control as the civilian head of the colonial Government in India. Curzon's political links, as well as his experience within the Indian subcontinent, usually ensured that his policy imperatives were implemented in a timely manner.

Pivotal to this was his conception of 'buffer states' that were notionally independent of British control but yet linked through a treaty relationship with the British colonial government in India. The implementation of this policy approach is briefly discussed hereunder with reference to Afghanistan, Tibet and the Persian Gulf/Middle East regions.

The Afghan question – neutralizing Russian aspirations for regional dominance

The Indian magazine *The Week* dated 30 January 2005 carried a series entitled 'Himalayan Military Blunders' dealing with lessons from past military failures. Possibly the most humiliating such policy blunder, with military overtones, was witnessed during the tenure of Lord Auckland when the British Resident, Sir Louis Cavagnari, and his staff were massacred in an Afghan uprising in Kabul in late 1842. Only one survivor, William Bryden, who reached Jalalabad on 13 January 1843, remained from an original British force of over 14,000. The British obsession with neutralizing Russian dominance by establishing dominance over Afghanistan resulted in at least three full-scale Afghan wars with multiple casualties on all sides. The 'Great Game', as Rudyard Kipling termed it, continued between Russia and England through most of the nineteenth century and vestiges of this rivalry are even now visible in modern day Afghanistan though the nomenclature and profile of the nation states concerned has changed.

The Curzonian approach conceived of strengthening Afghanistan as a strong buffer state enabling it to become a viable source of protection for the British colonial edifice in India. One component of this required strengthening of the defences on the north-west frontier of the then Indian border enabling them to absorb any kind of incursions from the Afghan side. This goal was successfully carried forward in Curzon's tenure and his efforts to strengthen the fort defences paid rich dividends in checkmating perceived Russian designs, if any, in this region. Curzon's arrival in India in January

1899 was preceded by the Pathan revolt (1897) and the revival of the 'Forward Policy' as a policy instrument within British colonial politics. The comments of Lord Roberts are particularly significant in this policy context. In a speech to the House of Lords in March 1898, he contended that the frontier region should be removed from the sphere of party politics owing to its overriding strategic dimensions. He stated that:

> [t]he Forward Policy, with its goal of extending influence and maintaining law and order on a part of the border where anarchy and murder have reigned supreme, is necessitated by the incontrovertible fact that a great Military Power is now within striking distance of our Indian possessions and in immediate contact with a state for the integrity of which we have made ourselves responsible. The Forward Policy must be pursued until the British colonial state establishes its dominance over the North West Frontier region.

Curzon had travelled extensively in Asia prior to coming to India and opined in his book *Russia in Central Asia* (1889) that the 'Russian interest is not Calcutta but Constantinople, not the Ganges but the Golden Horn. The sum and substance of Russian policy is to keep England quiet in Europe by keeping her employed in Asia.'

His solution for managing the troubled northern border was the creation of a new North West Frontier Province (NWFP) with headquarters in the city of Peshawar. In the words of Sir Olaf Caroe,

> The creation of the NWFP was recognition of the Pathan concept of oneness.

Curzon's decision provided an avenue for the recognition of Pathan self-esteem and also helped build and consolidate a firm national frontier, which survived World War I (1914–18) and only became subject to incremental fragmentation pursuant to the third Indo-Afghan war of 1919.

The Tibet issue – dealing with the spectre of Russian influence

Curzon's travels in Central Asia prior to assuming the post of Viceroy had convinced him that Russia posed the gravest threat to British colonial holdings in the Indian subcontinent. Hence his efforts to extend the idea of 'buffer zones' – whether notionally independent like Afghanistan or under British control like the NWFP – to India's strategic Himalayan neighbourhood in a proactive but composite manner. Information that Russia was keen to establish a closer relationship with Tibet, either bypassing or with the acquiescence of China, troubled Curzon and the then Balfour Government in London. The fact that this information was deduced more from innuendo and less from substantive intelligence mattered little to Curzon who felt that any perceived weakness would impact on the security of the north-eastern frontier of India. He often quoted the lessons of the Russian advance in the Pamirs, which had necessitated a stronger fortified frontier to be built as a reactive crisis response. Similarly he contended in January 1903 that any movement by Russia into Tibet would require a more strongly fortified north-eastern frontier in India. His comments, in a letter dated 12 March 1903 to the then Secretary of State for India, Hamilton, make interesting reading. He stated:

> If we are not to defend our own frontiers, to ward off gratuitous menace, to maintain our influence in regions where no hostile influence has ever yet appeared, until the national honour has been grossly affronted, the practical result will be that you will be unable to take a step upon your frontiers until they have actually been crossed by the forces of the enemy.

Curzon ultimately obtained approval from the Balfour Government for an expedition to be sent to Tibet under the leadership of Francis Younghusband. Curzon claimed that his goal was to neutralize nascent Russian influence on India's Himalayan borders while eschewing any desire to occupy Tibet permanently. Poor communications as well as ambiguity over British long-term foreign policy goals enabled Younghusband to exercise his own initiative in determining the terms of the Lhasa Convention signed on 7 September 1903. The time period of the Convention – 75 years – worried Curzon who nevertheless recommended its acceptance. The Balfour Government remained unhappy since they felt that the terms went beyond the original goals of the expedition by getting Britain entangled in Tibet without any commensurate tangible economic benefit to the then British Empire, except the possible neutralization of future Russian influence in the upper reaches of the Himalayan region.

Solidifying British dominance in Persia and the Middle East

The Curzonian paradigm of proactive policy formulation and implementation to neutralize his perception of any Russian threat to British colonial dominance even extended to the Persian Gulf region. This motivated his visit to Kuwait and other parts of the region in 1903. However, the emerging conflict between Russian and Japanese political and economic interests soon assumed centre-stage and resulted in the Russo-Japanese War of 1905. The Japanese victory served as a permanent dampener to Russian colonial ambitions and indirectly served to strengthen the Himalayan borders of colonial British India. The credit for focusing attention on the need for a comprehensive yet proactive policy on India's Himalayan frontiers and also implementing parts of his vision belongs to Curzon. His paradigm of proactive neutralization of efforts by other European powers to expand their influence in the region as well as the development of a string of buffer states to protect strategic British interests was both insightful and visionary. Despite this it is worthwhile here to mention that Curzon's vision for Britain at the beginning of the twentieth century was sometimes driven more by a sense of personal hubris than either foresight or conviction. Hence Edmund Burke's statement, reproduced below, in his monograph titled *Reflections on the Revolution in France* (1790), is relevant to comprehending the self-righteous rationale behind Curzon's policy motivations. Burke stated, 'I cannot conceive how any man can have brought himself to that pitch of presumption, to consider his country as nothing but carte blanche, upon which he may scribble whatever he pleases.'

Curzon's sense of personal righteousness and devotion to his policy goals remained visibly manifest in his policy imperatives throughout his stay in India as Viceroy and later in party politics in England. His inner motivations can be perceived from a personal undated note, later published in his papers, written by him about his work in India. He contended that he wanted to be judged by the test of results and stated: 'Great as may have been my errors, I had yet striven conscientiously and manfully for India and I cared not what the world might think or India say so long as I had this self-absorbing spring of conviction within me.' Perhaps the best analysis of Curzon as an individual came in the words of Sir Harold Nicholson (*The Spectator*, 4 April 1925), 'He loved and suffered with the eternal intensity of boyhood.'

The Himalayan frontier in post-independence India: strategic compromises and policy missteps

The dawn of the independent Indian nation state on 15 August 1947 was an event of substantive international relevance. The progressive dismantling of the British colonial edifice in the post-1945 world provided the impetus for the newly independent nations to re-evaluate their foreign policy priorities. Independent India under Jawaharlal Nehru opted for a policy of non-alignment with any power bloc and peaceful coexistence based on mutual respect. The 1917 Bolshevik revolution in Russia had spawned the new Union of Soviet Socialist Republics (USSR) whose foreign policy priorities in its immediate geographic neighbourhood were mixed. Similarly the Chinese revolution of 1949 led to the replacement of the erstwhile Kuomintang by the Chinese Communist Party led by Mao Zedong. India's strategic vision, articulated by Nehru, postulated coexistence dependent on a philosophical value-based policy construct rather than on the strength normally associated with the availability of a strong and effective military force. The efficacy of this approach was severely tested during the Cold War where identification and/or closeness with either of the power blocs – the American or the Soviet – were sometimes considered synonymous with state identity. It is to Nehru's credit that Indian foreign policy was able to divorce itself from these policy compulsions and instead devise a normative foreign policy based on equidistance from both the then American and Soviet power blocs. Unfortunately this paradigm, while solidifying India's moral stature, did little to protect its Himalayan borders with both China and the new Pakistan nation state that came into being in August 1947.

The dispute with China revolved around issues related to the McMahon Line, Tibet and India's borders with China. The post-1949 Chinese Government's desire to reassert its control over the Tibet region was as manifest in the actions taken to marginalize the Dalai Lama as it was to erode the significance of Buddhist religious practices. The Chinese denunciation of the McMahon Line and claims on large sections of Indian territory in Ladakh as well as the north-east (now Arunachal Pradesh) became a festering problem from the early 1950s onwards. Nehru's efforts to resolve boundary issues through bilateral dialogue with Chinese Premier Chou en-Lai proved to be a non-starter. Even India's high profile role within the non-aligned movement proved irrelevant so far as the border issue with China was concerned. In hindsight the reactive approach to the Himalayan border issue with the Chinese followed by the Nehru Government remained an unworkable policy construct. It allowed time for the Chinese to prepare for a confrontation with the Indian army, a reality that should have been anticipated after the Dalai Lama arrived as an exile in India in 1959. The exigencies of 'real politik' dictated caution but the approach of the Nehru Government demonstrated both weakness and an inability to debunk dialogue for strong pre-emptive action. The resulting policy construct spelt appeasement and emboldened the Chinese to attack the Indian border on the north-east and north-west in October 1962. By the time the ceasefire came, India's strategic Himalayan border interests had been reduced by the loss of valuable territory both in the Ladakh and then the North East Frontier Agency (NEFA) – as mentioned earlier, now Arunachal Pradesh – regions of India. It also denoted the strength of force as a form of deterrence over that of moral values and ultimately provided the fillip to long awaited Indian military reorganization and strategic policy overhaul.

The Chinese incursion was in the words of B. N. Mullik, 'a betrayal of trust' (*My Years with Nehru: The Chinese Betrayal*, 1971). However, its fallout remained more mixed. There was widespread criticism for a lack of foresight and an insistence that trust between nation states should never be one-sided, but reciprocal. India–China relations went into deep freeze thereafter and are only now slowly returning to a semblance of stability and reciprocal interaction.

The question of securing the Himalayan border adjoining Jammu and Kashmir also acquired prominence after the state's formal accession to India in October 1947. The induction of the Indian army into the state was effective in stemming the Pakistani advance and rolling it back. Once again, unfortunately, the compromise option was exercised and Indian policy makers led by Nehru thought that the breach with Pakistan was temporary in nature and could be resolved with outside mediation. Hence the decision by the Nehru Government to take the issue to the United Nations for resolution, a policy option which proved detrimental to India's long-term strategic interests.

Moreover, the upper reaches of the Himalayan border range remained in Pakistani control, which they later shared with the Chinese as a policy stratagem to resolve their own bilateral issues. The policy yardstick of appeasement as opposed to strategic confrontation influenced Indian foreign policy through the Nehru years. It is tragic that the confusion over a clear delimited border was never resolved leading to another potential policy misstep in Indo-Pak relations, especially in the upper reaches of the Himalayan region. Pakistani interest in returning Indian territory remained patchy at best. Hence the unilateral Indian decision to garrison the Siachen region in the late 1980s was strategically timely and significant.

The incipient conflict with China demonstrated to the independent Indian state apparatus that effective conflict management sometimes required the debunking of compromise in favour of force as a viable policy option. This ultimately happened in June 1999 when the Kargil incursion by Pakistan was repelled by the use of air and ground forces by the Indian Government. The inclusion of force as a viable policy imperative demonstrated strategic foresight and clarity in today's rapidly changing global scenario.

Strategic frontier issues in the twenty-first century – visible challenges and incipient opportunities

The end of the Cold War provided new strategic challenges for nation states in the post-colonial world. In India's strategic Himalayan neighbourhood this was manifested in reciprocal moves towards dialogue between India and China as well as between India and Pakistan. The placement of boundary issues for discussion between India and China was itself a strategic recognition of each other's potential, though a final solution is still far away and elusive.

The Pakistan issue is one that eludes resolution. The Indian control over Jammu and Kashmir, as well as Siachen, enables the nation state to place its strategic interests in the Kashmir region on a higher pedestal. Pakistan's control over Pak(istan) Occupied Kashmir (PoK) as well as Himalayan border regions like Gilgit, Hunza, Baltistan, Chitral etc., is unacceptable, though the option for a negotiated resolution in preference to force can be pursued. A coherent vision in the Curzonian mould is yet to evolve fully within the Indian strategic Himalayan policy scenario. But the process of crystallization has been initiated.

The post-September 2001 world has seen war and insecurity as direct end products of moves by nation states towards the eradication of terror as a policy option by non-state actors. Hence Central Asia is the new battleground in the 'War against Terror' and Afghanistan is enmeshed in its own internal squabbles in the nascent but emerging democratic mould. The developing challenges and opportunities available are briefly listed here.

The Curzonian paradigm of colonial expansion and creation of buffer zones is an invalid policy construct in the twenty-first century. The end of the Cold War, as well as the 'War on Terror', has spawned new political and economic challenges for nation states like India. Primarily among them are the following:

- emergence of new and potentially unstable state actors, especially in the Central Asian region;
- changing economic and political international alliances which makes the phenomenon of 'parallel diplomacy' in the Machiavellian mould necessary;
- evolution of old adversaries like China into stable nation states with a clearly discernible and coherent political identity, thereby necessitating a sustained strategy of strategic engagement through dialogue;
- arrival of multiple indigenous non-state actors with mixed political motivations but the ability and wherewithal to destabilize border areas both on the north and the east of India;
- expansion of non-state actor groupings in Nepal and Pakistan dedicated to violence, but with clearly defined political agendas focused on destablizing contiguous parts of India;
- the absence of an overarching and indigenous foreign policy construct which promotes the acquisition of military and economic muscle by the Indian state as a stablizing necessity and carries this philosophy into both bilateral and multilateral negotiations;
- accelerated and time-bound movement towards the aggressive utilization of proactive satellite imagery and the related state-of-the-art technology to protect India's strategic interests on its Himalayan frontier thereby enabling force deployment to be rationalized.

The international profile of the Indian nation state has certainly been enhanced positively after its emergence as a bona-fide nuclear power in May 1998. The structural dimensions of this reality can be multiplied by a proactive policy of engagement with nations straddling India's Himalayan borders thereby underlining the doctrine of pre-emptive anticipation rather than a reactive response construct. Pivotal to this is the need for quality intelligence gathering and interpretation at the human level especially at a time when non-state actors are fast emerging as key adversaries on India's Himalayan borders. Unfortunately the progress on remedying this shortcoming has been devoid of focus and patchy. The intelligence organization in India also needs to be revamped and reoriented much in the way the current Bush Administration in the United States of America (USA) has created a centralized intelligence organization integrating fifteen existing agencies. This remains an important variable for resolution in the overall policy framework which should evolve as a prelude to the development of an integrated Himalayan frontier policy for the Indian nation state in the twenty-first century.

Conclusion

An integrated policy paradigm for India's Himalayan frontier needs to be evolved expeditiously. This should articulate a strategic vision and also provide a menu of policy imperatives, which should be taken up for implementation on an incremental basis. The lack of such a policy construct makes reactivism rather than proactivism the defining norm and renders the ultimate policy paradigm incapable of responding to a fluid and constantly changing international situation. A case in point is the development in Nepal which led to the sudden removal of the elected democratic government and its replacement by an emergency regime headed by the former king on the perceived grounds of the government's inability in handling the Maoist insurgency in parts of the country. Such an event is of strategic significance for India's Himalayan borders and it is hoped that a future transparent but integrated policy construct will also provide a slew of available proactive options for Indian policy makers in advance of such events.

It is essential that the policy evolution process provides for interested citizens' input as a component of development. The institution of this approach would harness Indian minds and ideas in the goal of providing a viable policy construct for India's Himalayan frontier. Theodore Roosevelt said on 10 April 1899 that 'if we are to be really a great people, we must strive in good faith to play a great part in the world. We cannot avoid meeting great issues. All that we can determine for ourselves is whether we shall meet them well or ill.'

The visible challenges and available opportunities are before us as a society and the response has to be cooperative and integrated, with the government not playing just the role of thinker but also that of facilitator and implementer. Hence it is up to Indian society also to determine that this policy vacuum should be met by rising to the challenge in a proactive manner instead of waiting for reactive institutional responses to evolve. After all, in the words of Kemal Ataturk, 'A nation which makes the final sacrifice for life and freedom does not get beaten.' It is to be hoped that this proactive approach will also pervade the Indian body politic in the balance of the twenty-first century.

5 Strategic dimensions of the trans-Himalayan frontiers

Afsir Karim

Introduction

The great Himalayan range, the high Karakoram range and the Hindu Kush meet in the trans-Himalayan region and form a strategic triangle. Northern frontiers of British India rested on the high Karakoram range which formed the watershed between the Tarim Basin (in Xinjiang) and the Indus river system draining into the Indian Ocean. In the west, the frontier merged into the Pamirs and Hindu Kush mountain systems. In the east, the mountain frontiers ran into the high plateau of Tibet, which is bounded by Kun Lun in the north and the great Himalayan range in the south. The narrow wedge of the Wakhan corridor lies on the southern slope of the Pamir.

The trans-Himalayan region now encompasses Central Asia, Afghanistan, Xinjiang, Tibet and Jammu and Kashmir and this gives it a special strategic importance, making it an area of vital interest to America, Russia, China and India besides other countries within the region. There are two major entry routes into the subcontinent across the high watershed of the Karakoram range, in the east the Ladakh route and in the west the Gilgit route. Both the routes enter Ladakh and thereon lead to other parts of Jammu and Kashmir across high passes.

- The Ladakh route connects Khotan (Hotan), Yarkand (Shache) and Kashghar to Leh and other parts of Ladakh over the Karakoram range. Khardungla and Chhangla are the two major passes across the Ladakh range en route to Leh. The Zanskar range and the great Himalayan range still have to be crossed to reach the Kashmir valley.
- The Gilgit route enters Hunza across the Khunjerab and Mintaka passes leading to Gilgit-Abbottabad and thereafter to Pakistan's heartland. The Karakoram highway follows this route.

The strategic importance of Jammu and Kashmir can be well appreciated as two important land routes from the Tibet–Xinjiang region enter the subcontinent through its territory. The strategic significance of the trans-Himalayan region increased further after the arrival of Al Qaeda to this region and the installation of NATO and American-led coalition forces in Afghanistan. Central Asian countries have become the arena for the new 'Great Game' between the US, Russia and China.

Afghanistan and Tajikistan on the western edge of the region are neighbours of China, Pakistan and Jammu and Kashmir. The rugged terrain provides shelter to various ultra-radical Islamic groups and concealed routes of entry into Xinjiang from Pakistan and Central Asia. The fundamentalist groups, although subdued at present, continue their subversive activities without abeyance. The Xinjiang region remains a region of unrest due to the activities of the Muslim separatist groups.

The 'Great Game'

The northern frontiers of India were always a bone of contention between Britain, India and China. After the revival of Chinese nationalism in 1928, the new Chin–Shu–Jen administration of Sinkiang (now Xinjiang) adopted an aggressive posture on the boundary question. As the Chinese suspected the motive of the British Government, troops were despatched to the valleys of the Yarkand and Karakash rivers to watch the Karakoram watershed, with instructions to resist any intrusion by the British. A permanent garrison was stationed in Shahidulla under the Chief Administrator of Kashghar, and the Mir of Hunza was informed that the Chinese would henceforth tax his flocks, which had been so far traditionally allowed to graze north of the Karakoram watershed.[1]

In these circumstances, the British Consulate General at Kashghar was instructed to establish contact with the Kashghar authorities to forestall any new challenge to the traditional privileges of the Mir of Hunza. According to British perception, this situation in Sinkiang presented a new threat to the northern frontiers of India, especially after the collapse of the Chin–Shu–Jen regime and a Muslim rebellion that posed a threat to the regime in Sinkiang, which had now come under the control of Sheng Shih-ts'ai who was reported to be maintaining close contacts with the Soviet Union.[2]

In this environment, the British authorities in India feared that Bolshevik agents would infiltrate across the Karakoram range to spread their creed in India. By late 1890, it had become possible to travel directly from Rawalpindi to Gilgit via the Babusar pass, which was not a part of the Jammu and Kashmir state; the development of a direct route later prompted the British to negotiate the lease of Gilgit with the Kashmir Durbar. After prolonged negotiations in 1935, between the Government of India and the Jammu and Kashmir Durbar, it was decided that Gilgit Wazarat north of the Indus and its dependencies would be leased to the British for a period of 60 years and all civil and military administration would be transferred to the Government of India. It was, however, made clear that despite the lease agreement the area would remain an integral part of the Jammu and Kashmir state.[3]

Had India taken suitable measures after independence to safeguard the

Gilgit and Aksai Chin regions, Pakistan and China could not have occupied them and the Chinese could not have built the Tibet–Xinjiang link road across Aksai Chin. It seems both civil and military authorities in India were oblivious to the strategic importance of this region. Indian border claims were never strongly pressed with the Chinese Government, which India seemed to assume was settled by the exchange of the 1899 note between the British Government and China. In September 1959, Prime Minister Nehru wrote to the Chinese Prime Minister, Mr. Chou en-Lai, stating that 'The proposal made in 1899 referred not to the eastern frontier of Ladakh with Tibet but to the northern frontiers of Ladakh and Kashmir with Sinkiang. It was stated in that context that the northern boundary ran along the Kun Lun range to a point east of 80 degrees longitude, where it meets the eastern boundary of Ladakh. This signified beyond doubt that the whole of Aksai Chin lay in Indian territory.'[4]

India perhaps paid little attention to safeguarding the territory of the trans-Himalayan frontier because its main focus remained on the valley of Kashmir, ignoring their long-term strategic interests elsewhere in Jammu and Kashmir. If India had taken note of the fact that right from 1860 the British had started taking greater interest in the Ladakh region because of threats of intrusion across the northern frontiers over the Karakoram range, it would have taken appropriate steps to secure this area against Pakistani aggression in 1947.

Northern areas of Jammu and Kashmir: a survey

The northern frontier of Jammu and Kashmir, presently under the occupation of Pakistan, has great strategic importance and should be studied in detail by Indian analysts. This region comprises the high mountain region of the Gilgit–Skardu belt, which is sandwiched between the great Karakoram and the Hindu Kush ranges. The region shares common boundaries with Tibet, Afghanistan and Pakistan and has Tajikistan and Kyrgyzstan as next-door neighbours. Occupation of this area is of great strategic significance for Pakistan as it provides land routes to Xinjiang, Central Asia and the rest of Jammu and Kashmir. This region originally comprised a number of autonomous political and geographic entities that lay scattered over 27,188 sq. miles.

The British underscored the strategic significance of the region in 1868 when a separate Gilgit Agency was established to include areas where three famous mountain ranges, the north-western Himalayas, the great Karakoram and the Hindu Kush met. All these three ranges have their origin in the great Pamir Knot, although the north-western Himalayas are not directly linked with the Pamirs. Nanga Parbat and Hurmosh, two well-known Himalayan peaks, are located in this area and K2, the world's second highest peak, is located in this segment of the Karakoram range. Besides K2, at least 18 other peaks of more than 25,000 feet (7,620 m) are located in this segment.

Until 1892, Chitral formed part of the Gilgit Agency. Now, in the west, the boundary of Gilgit merges with Chitral, which is a part of the North West Frontier Province (NWFP) of Pakistan. On the eastern side, the boundary of Gilgit meets Xinjiang and Tibet along the Karakoram range. Across Mintaka, Shimshal, Khunjerab and other minor passes runs the Karakoram highway linking Xinjiang with Pakistan-occupied Kashmir (PoK) and Pakistan. In the north of Gilgit lies the Wakhan corridor of Afghanistan, at the junction of Afghanistan, Central Asia, PoK and Pakistan.

In 1928, the Gilgit Agency was divided into three parts:

- Gilgit Wazarat of Kashmir state, extending from Gilgit to Bunji and Astor;
- the districts of Hunza, Nagar, Punial, Yasin, Koh-i-Ghizar, Ishkoman and Chilas which for some strange reason was named 'The Republic of Chilas';
- unadministered areas of Darel, Tangir, Kandia (Killi) Jalkot, Sazin, Shatial and Harban.[5]

In the vast mountains of the Himalayas, the Karakoram and the Hindu Kush, in isolated valleys, cut off from the rest of the world, there is still habitation seeking a livelihood out of the remote mountains, and only occasionally crossing high passes during summer to venture beyond in search of new pastures or barter deals. The story of this hidden, little known region is nonetheless important as it comprises land of utmost strategic importance where Afghanistan, China, India, Pakistan and the Central Asian states converge.

The Pamir Knot occupies a pivotal position with respect to the Central Asian mountain system along with the Kun Lun, Alai and Tien Shan, which bifurcate into numerous subranges in the southern and central parts of China in the east, and northern and central parts of Afghanistan and Iran in the west. The north-western Himalayas and the Tibetan plateau have no direct link with the Pamirs but the Mustagh Ata, Kun Lun and the Karakoram, after emanating from the Pamir Knot, follow a south-eastern direction and merge with the Himalayas and the Tibetan Plateau in the east. The 'Main Mantle Thrust' in Pakistan and the geography of the Kohistan mass demarcate the northern boundary of the Himalayas.

Pak occupation of the Northern Areas

A fortnight before the partition (and independence) of India in 1947, the British terminated the lease of Gilgit and the region was given back to the Maharaja of Jammu and Kashmir. The Political Agent of Gilgit handed over his charge to Brigadier Ghansara Singh, a Governor appointed by the Maharaja.[6] Major William Brown, who was the Commandant of the Gilgit Scouts, volunteered to preside over the transition from the British administrators. However, it appears he had made up his mind in advance that Pakistan should be allowed to occupy this area as he had decided to serve in the Pakistan Army.

'Brown,' says a University of Cambridge website, 'and his second in command, Captain A. S. Mathieson, decided to use the Scouts to stage a coup d'etat and take complete control of the agency, and then offer it to Pakistan.' Whether Major Brown was acting on his own or on orders from above was not clear. Qutubuddin Aziz writes:

> An intrepid Scottish soldier, Major Brown was fond of Pakistan and hated the tyrannical Dogra satraps with Maharaja Hari Singh's evil coterie in Srinagar . . . The anti-Dogra rebellion in the Gilgit agency, in which pro-Pakistan Brown helped Hunza, Nagar, Punial and their neighbouring territories in the lofty Karakoram mountains to be placed under Pakistan's control in the autumn of 1947 which saw the birth of Pakistan.[7]

Since 1 August 1947, Brown had been advising the Governor of the Gilgit and Baltistan agencies, Ghansara Singh, and the Maharaja of Jammu and Kashmir that the 'correct course of action for Kashmir would be to join Pakistan'. He believed that the areas of Gilgit, Baltistan, Hunza, Nagar and Chitral should join Pakistan, and he made sure that this would happen.

The day following the accession of Jammu and Kashmir to India, Gilgit was declared as 'the independent Republic of Gilgit' which later opted to join Pakistan. The Gilgit Scouts and Muslim soldiers of the Jammu and Kashmir army were induced to assist in the occupation of Baltistan by Pakistan. In August 1947, a rumour was spread in Gilgit that the Government of Jammu and Kashmir was planning to disband the Gilgit Scouts. This led the 'disciplined and secular Scouts' to oppose the government. According to Pakistani sources, on 31 October 1947 Major Brown instructed the Scouts to take the Governor, Brigadier Ghansara Singh, into protective custody. Pakistan proclaimed that Brigadier Ghansara Singh was arrested on behalf of 'the independent Republic of Gilgit', and not on the dictates of Pakistan.

The so-called independent state of Gilgit lasted just 16 days and there is little doubt that Major Brown and Captain Mathieson were playing to the Pakistani tune. According to the *Daily Excelsior* of 1 November 1947, Major Brown declared that Hindu rule had come to an end and Gilgit would join Pakistan. The local leaders, however, favoured independence.[8] During this confusion, the Wali of Swat attacked Chilas and the Mehtar seized Koh-e-Ghizar and Yasin. The Gilgit Scouts, Chitral and Gilgit–Baltistan soldiers were soon organized into a new force.

Pakistan now administers the Northern Areas (NA) through a Commissioner who comes directly under the Ministry of Kashmir Affairs and Northern Areas. At present, the region is divided into the three administrative districts of Gilgit, Baltistan and Diamer, each under a Deputy Commissioner, with the respective district headquarters at Gilgit, Skardu and Chilas. The Baltistan Agency comprises Skardu Tehsil, originally a part of Ladakh district. The administration of Baltistan was put directly under the Government of Pakistan in 1961,[9] but some changes have been reported in this administrative set up lately.

The river Indus and its tributaries along with towering mountain peaks dominate the landscape of the Northern Areas. In this trans-Himalayan region, the main routes to the Kashmir valley cross various high passes. In undivided Kashmir, the so-called Gilgit transport route crossed from the Tragbal–Razdahnigam pass to Gurez and thence to the Kamri pass, Rattu Cant and finally to Astor (all in PoK now). This route is still of considerable strategic significance both to India and Pakistan. Another route from the Northern Areas to the Kashmir valley is via the Dras pass, now known as Zojila (its Ladakhi name); from Kargil a road originally led to Gultari and Olding and there is a bridge at Marol (all three in PoK).

In addition to the two routes mentioned above, there is a totally uninhabited tract through Deosai Plateau. It is the route that Pakistani forces took in 1947–48 to reach Dras and Zojila. The average height of this rugged area is between 12,000 and 13,000 feet, and is about 40 miles long. There is now a road over Deosai going to Skardu via Chilam Chowki, the Chachnoch pass – Barapam and Al Malik pass. There are two other routes from Astor to Skardu via Shegar Tham and Theraro, all mainly used by the Pakistan army. The centre of the plateau lies approximately 25 miles south of Skardu. There is a maze of mountains spread out in rough circular form ranging from 16,000 to 17,000 feet and within this ring there is the plateau, serrated by flat valleys, where the Shigar river joins the Dras river, which then flows into the Indus. Deosai is of glacial origin, and as a result this serrated but flat area was carved out. In the nineteenth century it was the most frequently used route between the Kashmir valley and Skardu. Durand, who explored this plateau in 1889, describes it in the following words:

> The Deosai plain is a great basin about forty miles across, averaging 12,000 feet in height and surrounded by a circle of snow rising three to four thousand feet higher. It is cut up by rolling spurs projecting from the main ranges into numerous broad shallow valleys through which run rapid streams. We were too early in crossing it, and the grass had not properly grown, though there were patches here and there and some flowers. It was a most desolate scene. The bare plains stretch away for miles without the vestige of a tree and with only here and there a few patches of stunted dwarfed juniper. Later on when grass springs up, they become favourite grazing grounds . . . the only animal life is furnished by the marmots.[10]

Astor is a famous Tehsil of the Chilas district. It has an old and famous fort and a grand mosque. The Darga Burzil and the Darga Rattu (two rivers) meet here to form the larger Astor river not far from the famous Rama Lake. Some years ago, Pratap bridge connected Bunji with Gilgit, but now a new alignment of the road and a new bridge leads to Gilgit via Bunji and Jaslok.

Hunza now forms a part of Gilgit district and retains its strategic importance due to its famous passes, prominent among them being Mintaka and Khunjerab. The majestic glacier of Rakaposhi adds to the mountain grandeur of Hunza. The inhabitants are mainly Dards in origin and are Ismaili Shias. They have been neglected and have never become reconciled to being a part of Pakistan.

Baltistan, as a whole, is a region of enormous mountains with heights ranging between 16,000 and 18,000 feet above sea level. The Shyok and Shigar rivers join the Indus to form a 20 mile crescent-shaped plain varying from one to five miles in width, south of Skardu. Skardu town is situated at the northern end of this valley at a height of 7,440 feet. In the early part of this century, Skardu was just a small village situated between a ruined Buddhist monastery and Kharpochu Fort. Durand described it as: 'Skardu, the capital of Baltistan or little Tibet is picturesque . . . it (the valley) is shut in by huge bare mountains with tremendous cliffs. We found Skardu possessed of an odious climate consisting of considerable heat in the daytime, and a gale of wind at night, which carries clouds of fine sand down the valley. It is said to be always windy here and consequently very cold in the winter.'[11] Skardu is now an important Pakistani airbase. The entire area has tremendous strategic importance for both India and Pakistan.

The importance of the trans-Himalayan region

The importance of the region to China and India is discussed below.

China

China's Xinjiang autonomous region has been a target of international *Jihadi* and Muslim separatist groups. Due to its proximity to Afghanistan, Pakistan and the Central Asian Republics allows them safe havens. *Jihadi* groups have easy access into Xinjiang from Afghanistan, Pakistan and PoK. Xinjiang is the hub of trade and interaction between China and Central Asia and provides routes of trade into Central Asia, India, Pakistan and Afghanistan. This area is, therefore, of immense strategic importance to China.

The Chinese have an abiding interest in the Central Asian region because of the natural resources and its geo-political importance. The Karakoram Highway that provides an essential link with Pakistan also opens the Indian Ocean region to China. Occupation of Aksai Chin has opened the gates of Central Asia, Pakistan and Afghanistan to China besides providing the vital link between Tibet and Xinjiang. Aksai Chin also provides easy routes of entry into Ladakh and can provide an additional link between the Northern

Areas of PoK and Tibet through the Shyok valley. In peaceful conditions, these areas can serve as good trade routes.

A peaceful border with India will help consolidation of the region by China as well as providing better connectivity with the Central Asian region. The opening of a gas pipeline from Xinjiang to India can provide new avenues of strategic and economic cooperation between India and China and development of trade and economic relations between the two countries will result in further opening up and development of the western regions of China. Cooperation in the trans-Himalayan region with India will also provide China with an opportunity to redefine its political and military ties with Pakistan in view of increasing dominance of the Pakistan–Afghanistan belt by the US and its NATO allies.

China would like to neutralize the increasing influence of the US and NATO in Central Asia and this is evident from increased Chinese efforts for cooperation with Russia to minimize and check spreading Western influence in this region. The agreements reached in the Shanghai Cooperation Organization (SCO) have helped in combating transnational *Jihadi* groups and protecting the national integrity and border security of China besides countering American domination. The SCO provides an inbuilt security apparatus against separatist organizations such as the *East Turkistan Islamic Movement* (ETIM). Another major advantage gained is related to checking of illicit drug trafficking, smuggling of weapons and movement of illegal immigrants into Xinjiang. The SCO is also the basic channel of multilateral cooperation that promotes the economic and security interests of China, Russia and Central Asia.

India

India has only a precarious hold on the strategic trans-Himalayan region since it lost control of the Northern Areas, PoK and Aksai Chin. Occupation of the Northern Areas by Pakistan denies India direct access to Afghanistan and Central Asia. Occupation of Aksai Chin and the eastern shoulder of the Karakoram pass has placed India at a great strategic disadvantage vis-à-vis China. Loss of strategic segments of the high Himalayas, parts of the Karakoram and opening of the Karakoram Highway in this region, marks a fundamental strategic shift in favour of Pakistan and China. This route along with the Chinese road running in Aksai Chin provides major avenues of collaboration against India by their own and Pakistani forces. The political and strategic dominance of passes in the Gilgit–Skardu belt in the west and the Karakoram pass and Aksai Chin in the east provide secure areas of influence and observation to Pakistan and China into Central Asia and Jammu and Kashmir. Pakistan has gained tremendous advantage by occupying what it calls the Northern Areas as this has enabled it to build the Karakoram Highway and establish a land link with the Chinese. This road link provided Pakistan means to instal complete missile systems from China undetected. By occupation of the Northern Areas, Pakistan has successfully cut off India's land routes to Afghanistan and the Central Asian Republics, and from this area Pakistan was able to mount successful operations against Zojila and Kargil in 1948. The routes of recent Kargil intrusions (1999) were basically the same. It needs to be emphasized that without Gilgit–Skardu in its possession, Pakistan cannot either surprise India or mount large-scale operations to destroy the Zojila–Leh road or pose a viable threat to the Shyok valley, Siachen and the Ladakh range.

It is evident that by denying important segments of the trans-Himalayan region to India, Pakistan has gained the strategic high ground, and India cannot change the status quo without a very expensive and prolonged military operation. A viable solution can emerge only through the ongoing peace process between India and Pakistan.

Current security environment

If the confrontation of India and Pakistan over Jammu and Kashmir continues, the strategic Northern Areas could become a bone of contention. The Aksai Chin problem between India and China is likely to remain unresolved at least in the near future and can prove a flash-point in certain circumstances.

The occupation of the Northern Areas by Pakistan and Aksai Chin by China continues to pose a major threat to India. The overland link between China and Pakistan through this area after the construction of the Karakoram Highway has added to the strategic significance of this region.

The Northern Areas can always be used by Pakistan to threaten the Zojila pass and the national highway between Srinagar and Leh. This area provides a perfectly concealed route of infiltration for Pakistani-sponsored irregular forces into Ladakh and Kargil, as was witnessed during the Kargil war in 1999.

The 'Great Game' has been revived after 9/11, as the subsequent developments in Afghanistan, Pakistan and Central Asia show. The Shanghai Cooperation Organization (SCO) founded in Shanghai in 2001, with Russia, China, Kazakhstan, Kyrgyzstan, Tajikistan and Uzbekistan as its members, is another manifestation of the 'Great Game' and points to the strategic importance of the Central Asian region. The main objective of the SCO is to strengthen strategic bonds between the member countries. The consultation includes political, economic and security cooperation in the area with a view to establishing peace and ensure stability. It is obvious that China played a leading role in establishing and strengthening the SCO for gaining a strategic foothold in this region and the trans-Himalayan region has gained added importance in this environment.

The Indian interest in Central Asia is mainly to develop trade and economic relations with the countries of Central Asia. India's attempt to achieve its aims is, however, hindered by India–Pakistan hostility. India suffers from a major disadvantage as the direct land routes through the Northern Areas to Central Asia and Afghanistan remain blocked for India. The alternative route to Central Asia through Xinjiang also remains closed because of the border dispute between India and China. The problem of terrorism and religious extremism which threatens the entire region could also have been addressed better if India had become a member of the SCO from its present status as an Observer Nation. India–China cooperation in Central Asia through the SCO in the context of their border dispute and counter-terrorism would have been extremely beneficial not only for both these countries but also for the entire region.

To establish viable overland commercial and industrial contacts with Afghanistan, Central Asian countries and Xinjiang, India has to create an amicable environment and establish friendly ties with both Pakistan and China.

The advent of NATO and the American-led coalition forces in the Afghanistan–Pakistan belt and the presence of Al Qaeda in the Waziristan–Northern Areas have added a new dangerous dimension to the political environment of the trans-Himalayan region. The unrest in Xinjiang and the India–China border remains unresolved. Aksai Chin area continues to be disputed and the Indian and Chinese forces face each other at close proximity in the region.

Uyghur separatist organizations active in Kazakhstan and Kyrgyzstan continue to maintain pressure in Xinjiang. The world Uygur movement is still pursuing subversive activities for destabilizing Xinjiang by abetting terrorist activities; pamphlets advocating *Jihad* against the Chinese have been frequently distributed in Xinjiang. The Eastern Turkistan Information Centre (ETIC) has been active in training militant groups for carrying out subversive activities; this group has made several attempts to blow up the important railway link between Lanzhou of Gansu province in China and Xinjiang.

The Islamist extremists and terrorists in Pakistan have of late become very active and they find safe havens in the mountainous areas of north-western Pakistan and the Northern Areas of Pok.

Taj.
Afghanistan
China
Territory ceded by Pakistan to China in 1963; claimed by India
Northern Areas
Siachen Glacier
Aksai Chin
Fed. Admin. Tribal Areas
Line of Control
Kargil
Line of Actual Control
Wular Lake
Srinagar
Vale of Kashmir
Jammu and Kashmir
POK
ISLAMABAD
Islamabad Capital Territory
Pakistan
Jammu
Himachal Pradesh
Punjab
India
China

Figure 6.1

6 The Ceasefire Line and Line of Control in Jammu and Kashmir

Evolution of a border

Sat Paul Sahni

The Line of Control (LoC) of today was envisaged as an international border in Jammu and Kashmir by the Western powers soon after Pakistan came into being in 1947. Over the years it has evolved as a border and has become a prospective solution of the Kashmir imbroglio. This paper is based on my personal knowledge as a war correspondent and information that I have gathered since 1947.

It is widely believed that India accepted the ceasefire in December 1948 docilely when in a few days the entire territory of Jammu and Kashmir would have been cleared of all the Pakistani invaders. This is not borne out by the facts. It has also been said that Sheikh Abdullah had 'forced' Pandit Jawaharlal Nehru not to liberate areas populated by non-Kashmiri speaking people. This is also a piece of fiction.

The die had been cast in the Western countries soon after the creation of the two Dominions of India and Pakistan. The armed invasion of Jammu and Kashmir territories gave these Western countries an opportunity to implement their plans. The manoeuvres and speeches made in the UN Security Council debates in the early months of 1948, by Philip Noel-Baker of Britain and Warren Austin of the US, indicate clearly that the Western countries were hell-bent on seeing that Indian troops did not liberate all the territory occupied by the Pakistani forces. On the ground, in Gilgit the British Commander of Gilgit Scouts engineered a revolt against the Maharaja's administration within eight days of the invasion in Kashmir.

In the UN Security Council, the US–UK combination with assistance from their allies were able to push through a resolution by a majority for appointing a five-member UN Commission (UNCIP) to go to the Indian subcontinent to 'place its good offices and mediation at the disposal of the Governments of India and Pakistan'. Its members were the US, Argentina, Belgium, Colombia and Czechoslovakia. Only Czechoslovakia was outside the orbit of, the Anglo-Saxon sphere of influence. UNCIP, soon after its arrival in the subcontinent, had requisitioned 40 military observers to be posted in the combat zone in anticipation of the ceasefire between the opposing forces and proposed the resolution which later became known as the '13 August 1948 Resolution'. Neither India nor Pakistan accepted it as proposed.

Both sought clarifications. But the Commission returned to New York and submitted its interim report to the Security Council. It did not draw any conclusions. Neither did it make any recommendations. Could it be a coincidence that on 22 November 1948, the UNCIP Report to the Security Council was not released in New York, but in Pakistan?

In the preceding months, the Indian troops had made significant advances. They had completed successful operations at Titwal, Uri, Gurez and Zoji La in Kashmir and Jhangar, Nowshera, Rajouri and Poonch in Jammu. They secured Kashmir valley from the north and established a land link with the Ladakh region, thus securing that vital area. In the Jammu region, Pakistan had been pushed back from its thrust towards the Akhnoor-Jammu city area, Rajouri had been liberated and the siege of Poonch lifted by establishing a road link with Jammu. The Indian Commander in the Kashmir valley, Major General K. S. Thimayya, had requested an additional brigade which would enable him to push Pakistan out of all state territories in Kashmir. Since the Indian Army was engaged in Hyderabad operations not even a battalion could be spared, Thimayya was told.

So after the middle of November 1998 no new operations were undertaken by the Indian Army. Since British officers in the Indian Army, Air Force and Navy were playing games which did not fit in with Indian plans, a decision was taken to replace the Army Chief, Gen. Sir Roy Bucher, by the most senior Indian officer, Army Commander Lt. Gen. K. M. Cariappa. The changeover date was to be 15 January 1949.

In view of the above developments the Western powers, using the forum of UNCIP, pressurized the Indian Government to accept a ceasefire at the earliest opportunity. They managed to get the Security Council to resume its debate on Kashmir on 28 November 1948 – only three days after the liberation of Kargil. The Council appealed to India and Pakistan to stop fighting in Kashmir and UNCIP was sent back to the subcontinent to make India relent. Dr. Alfred Lazano, accompanied by his assistant Mr. Sampier and the personal representative of the UN Secretary General, Mr. Colban, met the Indian Prime Minister Nehru in New Delhi on 20 and 22 December 1948. The Indian premier put forth doubts and apprehensions the Indian Government had about the Commission's proposals on ceasefire, truce and plebiscite. Dr. Lazano gave clarifications and assurances. These were put in writing and later became known as the 'Aides Memoire dated 21 and 22 December 1948'. India was thus being pressured to accept the Commission's proposals. It is interesting to see what was going on behind the scenes.

The Commission members were taking directions from their own governments and passing back the proposals before they were forwarded to New Delhi and Karachi. The Commission's report to the Security Council on 12 December 1948 was signed by only four of its members – Argentina, Belgium, Columbia and the US. The fifth member, Dr. Oldrich Chyle from Czechoslovakia, submitted a minority report which revealed the hidden hand. The Chyle report said, 'The failure of the Commission's Mission is therefore not solely ascribable to the intransigence of the Government of India and Pakistan. The reasons must also be looked for in the activities of the Commission per se with proper light thrown on the substance of the whole problem of Kashmir.' This report further revealed, 'The secret arbitration offer of the Commission was, before being presented to the Governments of India and Pakistan, placed at the disposal of the Governments of the US and the UK, and President Truman and Prime Minister Atlee, in synchronized action, made use of it for public interventionary pressure' (which came about on 31 August 1949).

Lord Birdwood in his book *Two Nations and Kashmir* has confirmed these reports. He says, 'the ceasefire between India and Pakistan was not as much an achievement of the Commission (UNCIP) as of the British initiative in the matter'. Later he refers to the role Gen. Bucher played in persuading Nehru to agree to a ceasefire. 'The Pakistanis were persuaded to a ceasefire because the Indian Army was threatening to drive Pakistan out of the [S]tate.'

When General Thimayya was refused additional forces to clear the remaining territories of Jammu and Kashmir state of Pakistani occupation, and India agreed to a ceasefire, the general proceeded on leave in the third week of December 1948 to express his unhappiness. Gen. Roy Bucher, the British Chief of the Indian Army, entrusted command of Indian troops to Brig. Henderson-Brook (who coincidentally was an Anglo-Indian). At the New Year celebrations in Srinagar's Amar Singh Club at midnight on 31 December 1948, the brigadier announced the ceasefire accepted by the Indian Army Commander-in-Chief to the assembled revellers. The ceasefire came into effect from the early morning of 1 January 1949.

It took some months to have an agreed Ceasefire Line marked and delineated on detailed maps. The Military Conference for this purpose was organized by the Commission at Karachi. It was held from 18 to 27 July 1949. The Indian delegation was led by Army Commander Lt. Gen. J. M. Shrinegesh, who was assisted by Major Gen. K. S. Thimayya and Brigadier S. J. S. Manekshaw. The Pakistani delegation was headed by the British Deputy Chief of General Staff, Major General W. J. Cawthorn. The UN Commission under whose aegis the Conference was held, was represented by its Chairman, Mr. Hernando Sampier (Columbia) and Mr. Williams of the US assisted by its Belgian Military Advisor, Lt Gen. Delvoi. Next day, the UN general surprised the Indian delegation by presenting a predetermined Ceasefire Line which was found to have been based on the Pakistani claim of territory under their control on 31 December 1948. The Indian delegation pointed out that the Commission's terms of reference stated clearly that it had to help the two sides to reach an agreement and not arbitrate between them. It took some hard arguments before this was accepted. After thorough talks supported by evidence and ground realities by the Indian delegation, an agreed Ceasefire Line was decided upon and marked on large size map sheets. The agreement was signed on the night of 27 July 1949.

The Ceasefire Line, according to this agreement, started from Manawar near

Akhnoor in Jammu region, moved westward, then skirted Poonch town on the west, went north, crossing the Jhelum river near Urusa (some distance away from Uri town in Kashmir). It moved to Chhota Kazinag and on to the Tutmari Gali. From there it went on to Richmaar Gali and thence north to the Kishenganga river. After Keran it kept going north along the river to Anzbari and thence to Burzil Nullah from where it went east to Karobal Gali. After Marpo La it went to Dalunang and further east through Gunderman and on to Point 13620 just overlooking Kargil town. Thereafter it continued through Chorbat La to Chalunka and then to Khor from where it went north to Glaciers. The Ceasefire Line ended at Point NJ 9842 near Thang.

Taking no decision to delineate the line beyond to the international border created tremendous difficulties in years to follow especially around Siachen Glacier. However, the Indian delegation was able to secure a major area of the Lolab valley of over 400 sq. miles (1,036 km^2) (north-west of Srinagar) and the Tilel valley – over 200 sq. miles (518 km^2) (north-west of Srinagar). These two areas were under occupation of neither army on the crucial night of the ceasefire. To monitor the ceasefire, the UNCIP had requisitioned 37 military observers from 1 June 1949 and an additional 24 observers from 24 June 1949.

From 1949 to 1965, both sides consolidated their positions on their own side of the Ceasefire Line. However, Pakistan continued to plan to unhinge India from Kashmir, politically and through subversion. It took advantage of Sheikh Abdullah's dismissal and of the situation following the disappearance of the sacred Relic of the Holy Prophet under suspicious circumstances from the Hazratbal shrine. During these 15 years, the Western powers persisted with efforts in the UN and elsewhere through plans put forward by Dixon, Graham and McNaughton Plans to achieve a solution around a division of Jammu and Kashmir. Outside the UN, the Western military blocs, SEATO and CENTO were used to pressurize India. In 1962, the situation arising out of the Chinese attack against India was used to persuade India to give concessions on Kashmir, in exchange for support and assistance.

In 1965, Pakistan sent infiltrators into Kashmir hoping to instigate Kashmiris to pick up arms. This was followed by full-scale attack by its armed forces. In this war India was able to liberate from Pakistani occupation nearly 25 sq. miles (65 km^2) in the Haji Pir sector, around 190 sq. miles (74 km^2) in the Chhamb sector and 20 sq. miles (52 km^2) in the Kargil sector. These were all strategic areas for the defence of Jammu and Kashmir. Although the UN Security Council brokered a ceasefire after 22 days of war, it took more than three months before Indian and Pakistani leaders met at Tashkent at the invitation of the Soviet Union. On 1 January 1966 India was asked to return all the liberated areas to Pakistan. This shocked the Indian Prime Minister. However, both sides agreed to 'observe the ceasefire terms on the Ceasefire Line'. The Western powers had managed to secure the Soviet Union's support for the Ceasefire Line as a de facto border in Jammu and Kashmir.

But five years later, in 1971, Pakistan again attacked India. Simultaneously all airfields in Jammu and Kashmir, along with many others in the Punjab and nearby areas, came under attack. Vicious attacks were launched in many other sectors, but fast moving Indian forces liberated large areas in the Turtok region in the Nubra valley of Ladakh. They also liberated the strategic hill-top overlooking Kargil town and many other areas.

Under the Shimla Agreement on 2 July 1972 all areas liberated by the Indian Army were returned to Pakistani control except 434 sq. miles (1,125 km^2) in the Kargil and Turtok sectors of Ladakh.

Article 4 (ii) of the Shimla Agreement deals with the Line of Control (LoC) as follows:

> In Jammu and Kashmir, the line of control resulting from the ceasefire of 17 December 1971 shall be respected by both sides without prejudice to the recognized position of either side. Neither side shall seek to alter it unilaterally, irrespective of mutual differences and legal interpretations. Both sides further undertake to refrain from the threat of use of force in violation of this line.

The Shimla Agreement was approved by the parliaments of the two countries. Subsequently, their military delegations met to delineate and demarcate the LoC. In November–December 1972 these meetings stretched over 12 days. The Indian delegation was led by Lt. Gen. P. S. Bhagat, the Army Commander, Northern Command. The two delegations signed a detailed set of 19 maps on which agreement was reached. Pakistan confirmed its commitment to the LoC on 11 December 1972 at Rawalpindi. It also said that India had claimed that it had seized 434 sq. miles (see ante), of Pakistani territory during the war, but according to the line now delineated, it had 341 sq. miles (883 km^2). Pakistan's claim was 59 sq. miles (53 km^2) on the Indian side of the Ceasefire Line and Pakistan retained all that territory.

From the mid-1970s, Pakistan quietly made plans to establish a claim over areas of the Siachen Glaciers beyond the Soltoro range of eastern Karakoram. They started with granting official permission to foreign mountaineering expeditions to climb peaks on the eastern side of the Siachen Glacier thus telling the world that they controlled the glacier. In April 1984, the Indian Army surprised them by moving into the area and occuping some of the high passes on the Soltoro range. While it established India firmly in the area, it also started a new confrontation making it the highest battlefield, and also the costliest. Pakistan contended that the LoC should run from Point NJ 9842 due north-east directly to the Karakoram pass. This would give it not only the whole of Siachen Glacier but also a common border for hundreds of miles with the Xinjiang province of China.

Since 1989, Pakistan has been engaged in a proxy war in Jammu and Kashmir. It wanted to do away with the LoC altogether and force India to quit Jammu and Kashmir through aggression from outside and uprising from within. It failed in its endeavour. During this period some elements in the US administration unabashedly lent their overt and covert support to Pakistan.

In 1999, the Pakistan army once again made a serious attempt to alter part of the LoC by occupying areas of Indian territory and cut off the Kashmir valley's land link with the Ladakh region. Though this misadventure took a heavy toll of Indian soldiers and the civilian population, it proved to be a total failure for Pakistan. It not only resulted in the death of hundreds of soldiers mostly from the Gilgit–Baltistan area, but also made Pakistan lose face before the international community. The US called its Prime Minister Nawaz Sharif to Washington where President Clinton asked him to withdraw all its soldiers and irregular intruders from the Indian side of the LoC in the Kargil sector. What is pertinent here is that not only the US but also all the important and major countries around the world asked Pakistan never to violate the LoC again. Pakistan was asked to observe the sanctity of this line scrupulously.

Concerned about cross-border terrorism, India over and again appealed, advised and demanded that Pakistan stop allowing infiltrating terrorists to cross the LoC. After 9/11, the US and many other countries have come to realize that Pakistan has to end cross-border terrorism and ensure that the LoC is not violated.

On the international border from Lakhanpur to Akhnoor (Manawar), on a 180 km stretch there is double fencing – both overground and underground. It is electrified at night and any human being or animal weighing 25 kg or more gets electrocuted if they step on the fence. During the fencing work Pakistan kept firing and even shelling to disrupt the work. At many selected stretches of the LoC there is double fencing as well, while in other places single fencing has been erected. Fencing has covered almost the entire length of the 734 km long LoC. It is estimated that erection of the fence has cost nearly 3 million rupees per kilometre on average. At strategic places along the fence, the army has installed ground sensors, battlefield surveillance radars, infrared sensors, and long-range reconnaissance and observation systems, hand-held imagers have also been co-opted. After initial objections and firing and shelling from across the LoC (which proved to be unsuccessful) Pakistan gradually reduced its objections.

It seemed the Pakistan army was being prepared to accept the inevitable. According to reliable reports Pakistan, on its side, has constructed many strong concrete bunkers all along the LoC, and strengthened and further fortified its formation headquarters and base close to the border.

One should draw attention to the geo-political importance and strategic configuration of the Gilgit–Baltistan region which Pakistan insists on calling its Northern Areas. It was in 1840 that the Dogra General, Zorawar Singh, conquered Baltistan and in 1846 that Karim Khan, a contender to the Gilgit throne, sought assistance from the Sikh Governor of Kashmir. An expeditionary force was sent from Kashmir which ultimately got Karim Khan enthroned but he was asked to rule in the name of Sikh Durbar. Through the Treaty of Amritsar 1846, the British transferred Kashmir territory including areas around Gilgit, which had been transferred to them by the Sikh Durbar after its defeat by the British. Soon after, there was an uprising in Gilgit. Maharaja Ranbir Singh despatched a large force in 1858 which suppressed it and brought peace and order to the area. The British soon realized that Gilgit was important for the defence of British interests in India, more so in view of the rising political ambitions of Tsarist Russia and their intended expansion southwards in Central Asia. The Kashmir Maharaja's assistance to the British on several occasions helped him to stave off British interference in Kashmir for many years. However, the British contrived to establish the Gilgit Agency and appointed a Political Agent in 1877. For some reason he was withdrawn but reposted there in 1881. By 1891, the British were able to bring the entire area of Gilgit, Hunza, Nagar and other principalities under their influence, until 1900 when they brought this entire area under their direct military control. The British, however, kept their pressure on Kashmir rulers for the next 35 years and in 1935 they forced Maharaja Hari Singh to grant to them an area of 1,300 sq. miles of Gilgit Agency on 60 years lease. It was on 1 August 1947, and only after the British withdrawal from India in that year, that the leased territory was handed back to the Kashmir Maharaja. A Dogra Governor, Brigadier Ghansara Singh, was appointed to administer the territory. The British cleverly left behind two British officers to command the local levy, the Gilgit Scouts. With local dissidents and pro-Pakistani elements, the Gilgit Scouts, led by Major Brown, rebelled and arrested the Dogra Governor on 31 October 1947 and took over the administration. On 4 November 1947, they raised the Pakistani flag and asked Pakistan to take over the administration. Within a fortnight, a new Pakistani Political Agent arrived.

Being unsure of its political status in this vital area, especially in view of the activities in the UN Security Council, Pakistan decided to play safe. It made the so-called Azad Kashmir administration and its political arm, the *Muslim Conference* transfer the responsibility and legal right over the Gilgit area. In Karachi an agreement was signed on 29 April 1949 between Pakistan's Federal Minister without portfolio, Mushtaq Ahmed Gurmani, on the one hand and the President of Muslim Conference, Choudhury Ghulam Abbas, the President of Azad Kashmir Government, Sardar Mohammad Ibrahim Khan, on the other. Matters handed over to Pakistan Government under the Karachi Agreement were:

- defence;
- foreign policy of Azad Kashmir;
- negotiations with UNCIP;
- publicity in foreign countries;
- coordination of arrangements of relief and rehabilitation of refugees;
- coordination of publicity in connection with plebiscite;
- all affairs of Gilgit, Ladakh under the control of the Political Agent.

In view of the Western intention of keeping the Gilgit areas out of control of India and also in view of the Karachi Agreement, UNCIP scrupulously avoided any mention of or decision in this area in any resolution or deliberations in its negotiation with India. India was intrigued and following the 13 August 1948 Resolution, Prime Minister Jawaharlal Nehru wrote a letter to the UNCIP Chairman on 20 August 1948. Among other matters, the letter sought explanation for the absence of any mention about the problem of administration or defence of the large area in the north. The letter also said that 'after Pakistan troops and irregulars have withdrawn from the territory, the responsibility for the administration of the evacuated area should revert to the Government of Jammu and Kashmir and for defence to us'.

On 25 August 1948, the Chairman of the Commission, Josef Korbel, wrote back to say, 'The Commission wishes me to confirm that, due to the [peculiar] condition of this area, it did not specifically deal with the military aspect of the problem in its Resolution of 13 August 1948. It believes however, that the question raised in your letter could be considered in the implementation of the resolution.' The Czech member of UNCIP, Dr. Chyle, in his minority report to the UN Security Council, pointedly said, 'The situation in the Northern Areas had meanwhile undergone a material change in that the Pakistan Army conquered many strategically important places during the interval. The Commission is bound to admit that while the reservation of the Government of India of 20 August 1948 may be legally valid, it is physically impossible to implement it.'

The Northern Areas have not figured in the territories of Pakistan as mentioned in the Constitutions of 1956, 1962 and 1973. In the Sino-Pakistan Agreement of 2 March 1962, Pakistan admitted that the Northern Areas are a part of Jammu and Kashmir. In this connection it will be pertinent to refer to the 'Azad Kashmir' High Court's Order on a petition presented in 1990 to it for determining the status of the Northern Areas. The Court ordered that since the Northern Areas were part of the undivided Jammu and Kashmir, these areas should come under control of the administration of 'Azad Kashmir'. Pakistan so far has not implemented this Order.

It is interesting to note that while the territory under the administration of 'Azad Kashmir' is only 13,528 km^2 the area comprising the Northern Areas is more than five times this area – or 72,495 km^2. Another portion of Indian territory was illegally ceded to China in 1962 by Pakistan. The Shaksgam valley and the Muztagh valley (east of Gilgit), an area of 5,180 km^2, was handed over to China as a sop when the latter was preparing to attack India.

Informed soundings were made on the settlement of a new border around the Ceasefire Line. According to the Union Home Secretary, L. P. Singh, at Tashkent in January 1966, Lal Bahadur Shastri reportedly gave his consent to former Soviet Premier Kosygin's proposal to President Ayub Khan that the then existing Ceasefire Line might be made the permanent India–Pakistan boundary with some adjustments but Kosygin reported that Ayub Khan had turned down the suggestion.

At the Indo-Pak summit at Shimla in July 1972, the Indian suggestion to change the nomenclature of the Ceasefire Line to the LoC (Line of Control) was accepted by Pakistan. In the words of Indira Gandhi's Secretary P. N. Dhar, 'the change was at the core of the Indian solution to the Kashmir problem. The de facto line was to be graduated to the level of de jure border.' We are told that the Pakistan Premier, Zulfikar Ali Bhutto, agreed to negotiate on the problem of Kashmir, provided that the solution, if arrived at, was implemented in a piecemeal manner and in step with the overall improvement in Indo-Pak ties. He was personally inclined to accept the status quo, that is, the LoC as a permanent solution. Mrs. Indira Gandhi was sympathetic to Bhutto's concerns over the attitude of hardliners in his army. Back home Bhutto justified his stance before the National Assembly and talked of the futility of war:

> 'We may not have gained anything by peaceful means, but how much have we gained by war?' he asked. 'The gain by war is that even we have lost half the country . . . We live in the same geographical area, we cannot get out of the subcontinent . . . Since we live here, let us find some way of living together, some *modus vivendi* . . . I think we will be able to bring peace to the people, peace that has been denied to them for centuries . . .'
>
> Events of the past 35 years make you ask, 'Did he mean it?'

Another dimension was added to the political imbroglio of Jammu and Kashmir after 9/11. The world started seeing reason in what India had been saying all along. India's fears and apprehensions of cross-border terrorism, the rise of fundamentalism and dangers of destabilization in Kashmir were better understood. The US and other countries seemed appreciative of the constraint and patience shown by India. Even in Pakistan, reasonable and saner elements started realizing the danger faced by Pakistan itself if it continued to pursue its old hate-India policy in Kashmir especially. The US stand on global terrorism and its armed intervention in Afghanistan and later in Iraq, started having an impact even on the Pakistan establishment.

American think tanks and influential persons, who are keen observers of South Asian affairs, openly airing their views on a Kashmir solution, have proposed the LoC as the new border in Jammu and Kashmir. Michael Krepon, President of the Henry L. Stimson Center of Washington, addressing a meeting at the Pakistan Institute of Strategic Studies, Islamabad in March 1999 said, 'Diplomatic efforts must take into account the reality of the LoC especially after the May 1998 nuclear tests by which the notion of either country moving to acquire fresh territory is unthinkable and the consequences would be unimaginable.' On 14 March 2000 in Washington, the US Secretary of State Madeleine Albright called upon both the countries to take 'tangible steps to respect the LoC'. On 18 March 2000 at its meeting in Berlin, the European Commission called on India and Pakistan to respect the LoC. This was followed by the then US President, Bill Clinton, at a press conference on 21 March 2000 in New Delhi asking both countries to 'respect the LoC', while reiterating his administration's position about the inviolability of the LoC. In London on 1 February 2000, the former British High Commissioner to New Delhi, Sir David Gore-Booth, called for recognizing the LoC as the international border as the 'only feasible solution to the Kashmir problem'. He further said, 'I am sure that this is what Zulfiqar Ali Bhutto and Indira Gandhi had in mind at Shimla in 1972' (when the two Prime Ministers signed the Shimla Agreement).

John H. Sandrock, the director of another important US think tank, the Atlantic Council of the International Security Programme told a joint group of visiting Indian and Pakistani journalists in Washington on 11 August 2004 that 'the LoC is going to become the international border one day'. The Washington based Carnegie Endowment for International Peace's Senior Associate Ashley J. Tellis wrote in November 2004 that Pakistan should seek a solution of the Kashmir problem 'without any substantial territorial change or radically altered sovereignty'.

For some years many Indians had been talking about the LoC as a possible solution but the first Indian politician who publicly proposed this as a likely border in Kashmir was none other than the former Chief Minister of Jammu and Kashmir, Dr. Farooq Abdullah. It was obvious that as a Kashmiri Muslim he was prepared to stake his political future on this proposal. While on a visit to the US in 1999, addressing a mixed audience of diplomats past and present, US officials, experts and journalists at Washington's Henry L. Stimson Center, Farooq Abdullah urged both countries to accept 'the LoC as the international border and work to make that border peaceful'. He asked Pakistan and India to be content with what they both have and let the issue be resolved for all time.

In the conferences of political leaders from both sides of the LoC, first at Toronto and then at Kathmandu in December 2004, even the separatists were coming round to considering that the status quo was the only guarantee for peace in Kashmir. They wanted both countries to give up rigidity for flexibility. And this would mean peace in Jammu and Kashmir.

The Indian case has been forcefully presented in the recent past. Dr. Farooq Abdullah, in December 2004, rejecting Musharraf's new plan of seven divisions of Jammu and Kashmir, stated that India will not allow any further division. The Prime Minister, Dr. Manmohan Singh, a few days later said that India would not allow another partition of the country or permit redrawing of boundaries but 'we are willing to look at various options'. He is reported to have told the Pakistani Premier, Shaukat Aziz, when he visited New Delhi on 24 November 2004 that the territorial status quo cannot be changed.

Field Marshal Sam Manekshaw who was involved with the finalization of Ceasefire Line in 1949 and again in 1972, when it was converted into the LoC, believes it should be turned into an international border. In a lecture delivered in Pune on 23 August 1999, Manekshaw argued that: 'If Pakistan-occupied Kashmir is inducted into the Indian Union then we will inherit one million people who have no loyalty to us. It will be the biggest problem that we would take on.' Another former Indian Army Chief, General Shanker

Roy Chowdhary, MP, during his lecture at Dehradun on 22 August 2000 said that he 'believed that the only practical solution of the Kashmir issue is to convert the Line of Control into an official border'. Former Prime Minister of India Chandra Shekhar also supported this. Speaking to the press on 18 May 2003 he stated that converting the LoC into an international border was 'the only solution to the Kashmir problem'.

In the recent past, thinking among Pakistani intellectuals also seems to be undergoing a change. Dr. Israr Ahmed writing in the Urdu daily *Jang* said, 'Pakistan cannot take Kashmir by force, nor will India offer it on a platter'. He advised the Pakistani Government to seek a solution in the spirit of give and take, because Pakistan was bound by the Shimla Agreement. Mr. Rasheed Rahman writing in the *Daily Times* advised Pakistan to accept reality and that is the status quo. Another writer, Mazir Naji, wrote in the same newspaper in January 2005, '[the] Kashmir solution will be no different than what Europe and US have already decided – with Indian approval'. A very perceptive and widely respected Pakistani analyst, Ayaz Amir, believes that there is going to be no military solution, only negotiations would be acceptable. 'If we don't get the Vale and if India does not agree to a plebiscite', the negotiation, according to him, will mean Pakistan offering unilateral concessions: 'no support to the Kashmiri freedom struggle'. He was of the view that 'for Kashmir's sake we should not imperil our own security or put on ourselves a burden we cannot carry'.[1]

The former Indian Minister for External Affairs Natwar Singh explained that there is no quick-fix solution and he likened the Indo-Pak peace process to a marathon and not a 100 metre race. Pakistan Premier Shaukat Aziz advocated a strategic rethink and bold decisions and said, 'We must be realistic and pragmatic.' The wheel seems to be going round full circle. But it will be a long arduous journey before it becomes an accepted reality.

7 The Northern Areas of Jammu and Kashmir

B. Raman

Introduction

While the movement launched by the Baloch nationalists in Balochistan has received the attention of the international community, similar attention has not been paid to the growing unrest in the Gilgit–Baltistan area of Jammu and Kashmir, which has been under Pakistani occupation since 1948. While the Pakistani authorities refer to this area as the Northern Areas of Pakistan, the local nationalists, who have launched a separatist struggle, call it by its historical name of Balawaristan. Before 1948, this area, which the then maharajah of Jammu and Kashmir had given on lease to the British since 1935 in order to enable the British to keep a watch on the developments in Xinjiang and Afghanistan, used to be known as the Northern Areas of Jammu and Kashmir. General Zia-ul-Haq, Pakistan's military dictator between 1977 and 1988, had it renamed as the Northern Areas of Pakistan, at the culmination of a process of integration of the territory into Pakistan.

This area, which borders on India, China and Afghanistan, has been of strategic concern and interest to India, Pakistan, China and the US. To India, because, first, it is its territory, which has been under the illegal occupation of Pakistan since 1948; second, this area, particularly Baltistan, has close ethnic, religious, cultural and other historic links with the Ladakh region of Jammu and Kashmir, of which it used to be a part before the Pakistani occupation; third, the Shias and the Ismailis of the area, who constitute the majority, have close fraternal links with the Shias of the Kargil area of Ladakh. They look up to India and its Shias for moral support in their struggle against the Pakistani authorities for the right of self-determination for the Shias of Pakistan in general and of the Northern Areas in particular. Fourth, part of the *Jihadi* terrorist training infrastructure of Pakistan's Inter-Services Intelligence (ISI) is located in this area; and, fifth, the growing Wahabization of the local population promoted by the Pakistan Army since the days of General Zia. The developments in this area have an important bearing on India's national security.

Pakistan is concerned because the river waters which sustain agriculture in the Punjab flow from this area. The Karakoram Highway from the Xinjiang province of China, constructed with Chinese help in the 1960s and the 1970s and inaugurated in 1978, is of tremendous strategic significance for the clandestine road transport of nuclear missiles and other military material from North Korea and China to Pakistan; the mountain heights in this area provide vantage points for any Pakistani attempt to cut off the Ladakh region from the rest of Jammu and Kashmir; and the Shias of this area have always proved to be a thorn in Pakistan's flesh. China's, concern is the Karakoram highway which helps it in keeping the Pakistani military strength sustained against India, which has always been an important Chinese strategic objective; the highway provides access to the sea for Chinese exports from the Xinjiang province through the Gwadar port on the Mekran coast in Balochistan now under construction with Chinese assistance; and because of the likely impact of the growth of Wahabi extremism in this area on the activities of Uighur nationalists and *Jihadi* terrorists in the Xinjiang province.

The area is strategic to the US in order to keep the pro-*azadi* (freedom) and not the pro-*Jihadi* separatist groups in Xinjiang sustained in their struggle for Uighur independence and to enable them to counter the activities of *Jihadi* terrorist groups; to keep a watch on the Chinese nuclear establishment located in the Xinjiang province; to use the local Shias, who are strongly against bin Laden, in its attempts to have him, his second-in-command Ayman al-Zawahiri, and other Al Qaeda survivors smoked out.

Since the 1980s, there has been a steadily growing ferment against Islamabad in the Northern Areas for the reasons explained below.

Changes in demographic composition

A conscious policy of bringing about a change in the demographic composition of the area was initiated by the late General Zia after the success of the Islamic Revolution in Iran in 1979, in order to counter the growing sectarian consciousness of the Shias and their demand for political and economic rights on a par with the Sunnis. This change was to be brought about by encouraging and facilitating the migration of Sunnis from the other provinces and the Federally-Administered Tribal Areas (FATA) and by resettling Sunni ex-servicemen in the area. As a result, the Shias and the Ismailis, who constituted about 85 per cent of the population in 1948, today constitute only about 53 per cent of the population. The Sunnis form about 42 per cent, and the remaining 5 per cent belong to other sects of Islam.

In 1948, the sons of the soil (Ladakhis, Baltis and related ethnic groups) constituted about 80 per cent of the total population. Today, they constitute about 53 per cent. Pashtuns from the North-West Frontier Province (NWFP) constitute about 24 per cent, Punjabis about 6 per cent, Mirpuris from the Pakistani Occupied Kashmir (PoK) about 3 per cent, Pashtuns from the FATA about 1 per cent, Sindhis and Mohajirs from Sindh about 1 per cent and Balochs less than 1 per cent. The provincial and ethnic origin of the remaining 11 per cent is not known.

Zia-ul-Haq not only encouraged and facilitated the migration of people from the other areas of Pakistan to the Northern Areas, but also assisted the anti-Shia Sunni extremist organization *Sipah-e-Sahaba Pakistan* (SSP), then known as the *Anjuman Sipah-e-Sahaba*, to set up its presence in the area and start a large number of *madrassas* (religious schools). The intent was to impart religious education to the local Sunnis in the Deobandi-Wahabi ideology and military training through the ex-servicemen in order to resist Shia militancy. Since the state's expenditure on education was negligible, the *madrassas* became the main centres for the education of local children and the breeding ground of extremism and terrorism. Due to the lack of economic development, the Northern Areas have the lowest GDP growth rate in Pakistan today, but the third highest GMP (Gross *Mulla* Product) and GJP (Gross *Jihadi* Terrorist Product) growth rates. The local Deobandi-Wahabi *madrassas* produce the third largest number of Mullas after the NWFP and Balochistan and the third largest number of *Jihadi* terrorists after Punjab and the NWFP.

The resentment of the Shias over the change in the demographic composition and the Army-sponsored induction of Wahabism and Sunni extremist elements into the area to counter the growing political activism of the Shias, allegedly funded by Iranian intelligence, created a sectarian divide between the Shias and the Sunnis in an area where no such divide existed before the days of Zia. The Shias of the Northern Areas, like their coreligionists in the Kargil area of the Ladakh Division, were known to be a peace-loving, tolerant people. The Sunni extremism led to the emergence of extremism in the Shia community encouraged by the *Tehrik-e-Jaffria Pakistan* (TJP) and its militant wing *Sipah Mohammad*. This also led in 1988 to demands from the Shias for the creation of an autonomous Shia state to be called the Karakoram state. Alarmed by the signs of militancy in the Shia community, Zia-ul-Haq put Pervez Musharraf in charge of the suppression of the Shias. Musharraf did so ruthlessly in 1988 and was helped in this by a group of tribesmen from the NWFP and the FATA.

After the death of Zia-ul-Haq in a plane crash in August 1988, the Northern Areas have seen frequent eruptions of Shia–Sunni clashes resulting in fatal casualties over issues such as the curriculum in the local schools, which excluded lessons on the beliefs of the Shias, discrimination against the Shias in the recruitment to government services etc. After violent riots in June 2004 over the question of the school syllabus, which the Shias viewed as anti-Shia, the *Daily Times*, the prestigious daily of Lahore, wrote as follows in an editorial titled, 'The plight of the Northern Areas':

> The syllabus issue in Gilgit continues to hang fire. On June 3, the local administration imposed [a] curfew in the city after clashes broke out between the police and the Shia mobs protesting the new syllabus. Earlier, a meeting between the leaders of the community and the administration had failed to break the impasse on the issue.

> What's going on?
>
> Sporadic news has been coming out of Gilgit about the syllabus problem for the past one year. We know that the Shias there are unhappy over certain passages and pictures in the officially prescribed Islamiyat textbook. But it is a measure of the failure of the mainstream press that most newspapers have not bothered to dig up the facts. Every story talks about the unrest and refers to the syllabus, but no attempt has been made to provide details of what exactly is wrong with the textbook.
>
> This is in line with the treatment we have meted out to the Northern Areas over the past four decades.
>
> In the 1980s, the Pakistani state under General Zia-ul-Haq made a deliberate attempt to infuse Sunni–Deobandi cadres of a sectarian party in order to put down the Shia. The problem of syllabus we now encounter in Gligit and also elsewhere is the product of the state's enterprise of backing a particular brand of Islamic exegesis. Of course, there are other factors, not least the rampant corruption in the region by officials of the state and lack of development and employment opportunities. But the ground reality is that the area is sitting atop a time bomb and the syllabus is the trigger that could activate it. That is why it is surprising that despite the issue festering for so long the federal government has done nothing visible to address it. Now it threatens to become a law and order problem.

Earlier, the violent activities of the Sunni extremists were directed only against the Shias, but not against the Ismailis. In 2004, the Sunni extremists started attacking the Ismailis and the schools run by the Aga Khan Foundation not only in the Northern Areas, but also in the adjoining Chitral area, in protest against the examination system followed by the Foundation.

The Aga Khan University Examination Board (AKU-EB) had been established through an ordinance to give the country an efficient system of exams that all students could afford. This was done in view of the growing popularity of the GCSE and 'A' level examinations conducted by the University of Cambridge throughout Pakistan. Each student taking these exams had to find and pay around Rs 20,000. After the Board's programme comes into force, an examinee would pay only Rs 1,500 if he came from a non-profit-making school and Rs 3,000 if he belonged to a private school. The standard of examination was supposed to be as high as that of Cambridge because it was reliable and recognized in the private sector.

It should be noted that the AKU-EB ordinance applies, so far, only to the private sector and the federal institutions and is completely voluntary. The Board projected it as secular and anti-Islam.

After the murder of two Ismailis in Chitral on 27 December 2004, the *Daily Times* wrote in an editorial entitled 'Chitral trouble is symptomatic of deeper malaise':

> Four masked men killed two workers of the Aga Khan Health Services Office in Chitral on Monday (27 December 2004) and burnt four vehicles

> belonging to the charity organization. The police have registered a case against the unknown assailants and have also arrested four persons belonging to a banned organization.[1]

This kind of violence has happened in the area before, but has gained momentum after the MMA campaign against the Aga Khan Foundation in the rest of the country. In the adjacent Northern Areas (Gilgit) the Aga Khan's charity institutions have come under attack regularly in the past few years after being targeted by the radical religious elements waging *Jihad* in Kashmir. In the early part of 2004, the Northern Areas witnessed sectarian unrest for almost six months. Schools were closed and there were instances of sporadic violence in areas where Shia and Ismaili populations were concentrated, but where power and influence have passed to Sunni clerics. In Chitral, the Shia–Sunni tension dates back to 1988 when the Northern Areas were attacked by Pushtun *lashkars*. That year General Zia-ul-Haq dismissed Prime Minister Mohammad Khan Junejo for failing to control violence. Today, the MMA (*Muttahida-Majlis-e-Amal*, a coalition of six fundamentalist parties) clerics are openly threatening 'action' against the Aga Khan Foundation because they do not want it to organize a better examination system in the country. There is no doubt that the fiery sermons delivered down-country are having their effect in the north and have also intensified sectarian conflict in Chitral.

The *Daily Times*, in its editorial, exposed the politics of MMA Opposition to the progressive education in Chitral:

> Although the MMA, led by *Jamaat-e-Islami*, has no past record of criticizing the Cambridge system in the country, the *Jamaat* now says the AKU-EB is set to 'secularize' the country by the introduction of this system. How is that possible through mere conduct of such exams? The ordinance establishing the AKU-EB says quite clearly: 'The Examination Board shall follow the national curriculum and syllabi'. There is no hidden reference here to any presumed secular brainwash as feared by the clergy. So what is the truth of the matter? The truth is that a hidden desire to exclude one more community from the pale of Islam persists after what the religious fanatics have done to non-Sunni majority locations in the North. What was happening so far in the periphery is now threatening to come to the centre. That is why General Pervez Musharraf must take firm action against the elements which have attacked the Aga Khan Health Services Office in Chitral and are working under a scheme to destabilize the country by exacerbating its sectarian conflict. That is also why he should seriously think of displacing the reactionary MMA with a liberal party in his political affections.

At least 14 people were killed, 6 of them burnt alive, and 14 injured during sectarian attacks in Gilgit on 8 January 2005, after which a curfew was imposed on the city and troops deployed to restore order. The clashes took place after 'unidentified' people shot at the car of Agha Ziauddin, a Shia community leader and *Imam* of the main Gilgit mosque, killing two of his bodyguards and seriously wounding him. One of the assailants was shot dead when fire was returned. Ziauddin succumbed to his injuries leading to more violence throughout Gilgit and Baltistan and the imposition of a curfew for nearly a fortnight. Ziauddin was in the forefront of the campaign against the school curriculum. The *Daily Times* wrote in another editorial as follows:[2]

> Following the incident in Chitral, the chief of the banned *Lashkar-e-Toiba*, Hafiz Saeed, proclaimed in Lahore that the government was 'apostatizing' the Muslims of the Northern Areas, meaning that it was supporting the so-called 'heresy' of Ismaili and Shia Islam. The *Lashkar-e-Toiba* gained influence in the Northern Areas during the Kargil Operation in 1999, not without causing some sectarian incidents. From being a completely Ismaili region in history, it has been injected with external populations through natural immigration from the rest of the country. But there have been manipulations too, as a result of which the region has suffered violence.
>
> Saturday's killing in Gilgit is a big incident recalling the 1988 massacre which accounted for 44 deaths after *lashkars* sent in by a politician nicknamed the 'devil of Hazara' entered the Shia city after travelling the Karakoram Highway which was supposed to be guarded closely by the Pakistan Army. Then it was the high tide of General Zia's *Jihad* in Afghanistan and the Shias – from Kurram Agency to the Northern Areas – were considered 'non-cooperative'. That year, Parachinar and Gilgit were both subjected to invasions and hundreds of people were put to death. The climax of the anti-Shia campaign was reached when the all-Pakistan Shia leader Allama Arif ul Hussaini – a Turi from Kurram Agency and close companion of Imam Khomeini – was murdered in Peshawar. Shockingly, ten days later General Zia was himself killed in an air-crash in Bahawalpur.
>
> Was the Musharraf government not forewarned? Sadly, it was, when last year there was unrest in the Balti Shia areas and the local population gathered several times in protest against the textbooks being prescribed in their schools. There were also complaints against clerics coming from 'outside' the area and delivering fiery sermons based on sectarian hatred. But nothing was done. The incidents were not treated as a series of connected happenings leading up to a climax. Islamabad seems to be more concerned about mollifying the clergy on 'religion entry' in the passports than about thinking of how to save our vulnerable populations from increasingly falling victim to religio-ideological policies.

The absence of democracy

The Northern Areas have the status of a Federally-Administered Area and are treated on a par with the Federally-Administered Tribal Areas (FATA).

The Chief Executive authority for the Northern Areas is vested in the Federal Minister for Kashmir Affairs, Northern Areas, States and Frontier Regions. He is assisted by a Deputy Chief Executive, who is appointed by the Chief Executive from amongst the members of the Northern Areas Legislative Council (NALC), after he has secured the support of the majority of the members of the Council. The Deputy Chief Executive enjoys the status of a Minister of State and is in turn assisted by advisors who are appointed by the Chief Executive, in consultation with the Deputy Chief Executive, from amongst the members of the NALC. The advisors have the status of a Provincial Minister. The Administration is headed by a Chief Secretary, and Secretaries head the departments. The Northern Areas Rules of Business, 1994 defines 'Government' as meaning the Chief Executive, the Deputy Chief Executive and the Chief Secretary, Northern Areas.

The Northern Areas Legislative Council is an elected body, having six members each elected from the three districts of Gilgit, Diamir and Baltistan, and three each from the two districts of Ghizar and Ghanche. There are five reserved seats for women, one from each district. The 24 directly elected representatives in the Council elect the women members for these seats. The Federal Minister for Kashmir Affairs and the Northern Areas is a member of the Council. The Council elects from amongst its members a Speaker, who is given the status of a Provincial Minister. Schedule II to the Northern Areas Council Legal Framework Order, 1994 lists the matters with respect to which the Council may make laws. No bill passed by the Council can become law unless it is approved and signed by the Chief Executive. The Government of Pakistan may also by order make laws with respect to matters not enumerated in Schedule II. The annual budget allocated to the Northern Areas is presented before the Council in the form of a statement.

In the rest of Pakistan, the allocation of federal resources to the provinces is made on the basis of population, with some weightage for the underdeveloped or backward areas. This principle is not applicable to budgetary allocations to the Northern Areas. The Northern Areas administration is treated as an attached department of the Federal Ministry of Kashmir Affairs and Northern Areas, which calculates the budgetary requirements of the PoK and the Northern Areas, gets them sanctioned by the Federal Finance Ministry and then allocates them to these two respective administrations.

It is the Chief Executive (Minister) who decides further allocation of funds. The Northern Areas Rules of Business, 1994 provide in Rule 5(c) that the Chief Executive shall exercise the powers of the Federal Ministry of Finance in relation to the approved budget for the Northern Areas and that he shall exercise administrative powers of the establishment division in relation to employees of the Northern Areas.

The Northern Areas are ruled from Islamabad by the Federal Minister for Kashmir Affairs and Northern Areas, who is designated as the Chief Executive. While the Deputy Chief Executive is generally from the area, the

Chief Secretary, the Inspector-General of Police and the heads of departments are generally from other provinces, mainly from the NWFP. The so-called Legislative Council has very little legislative power and such little financial power as it has is concentrated in the hands of the Islamabad-based Chief Executive. There is no provision for a leader of the opposition and for a vote of confidence. The only duty of the Legislative Council is to rubber-stamp the decisions of the Chief Executive and to carry out his wishes. There is no Public Service Commission to make recruitments to the local government services from amongst the local population.

There was no reference to the Northern Areas in the Pakistani constitutions of 1956, 1962, 1972 and 1973. Nor is there a reference to it in the 1974 Interim Constitution of PoK. The only reference to it is in a Legal Framework Order, which placed it under the control of the Kashmir Affairs and Northern Areas Ministry. In 1982, Zia-ul-Haq proclaimed that the people of the Northern Areas were Pakistanis and had nothing to do with the state of Jammu and Kashmir. Following this, some residents of the Northern Areas filed a writ petition under Section 44 of the PoK Interim Constitution Act of 1974 in which they challenged the Pakistani view that the Northern Areas were not a part of Jammu and Kashmir but a part of Pakistan. They also contended that even the Sino-Pakistan Agreement of 1963 had specified that the Northern Areas were a part of that. Rejecting their contention, the Government of Pakistan claimed that it 'was not functioning or operating within the territory of Azad Jammu and Kashmir (and) as such it was not amenable to the jurisdiction of this court'. It also denied the existence of the well-known Karachi Agreement of 28 April 1949 'whereby the administrative control of Northern Areas was delivered to the Government of Pakistan'. The High Court of PoK, however, ruled that the so-called Northern Areas were a part of PoK. Pakistan had the ruling vacated by the Supreme Court of PoK which said that the High Court had no jurisdiction to issue the ruling.

In another case when the Al Jihad Trust and others filed a petition before the Supreme Court of Pakistan demanding the granting of fundamental rights, including the right to vote and be represented in the Federal Parliament and the right of self-determination, the Government of Pakistan argued that the Supreme Court of Pakistan had no jurisdiction since the Northern Areas were not, in terms of Pakistan's constitution, a part of Pakistan. Subsequently, the Supreme Court of Pakistan in a landmark verdict on 28 May 1999, directed the government to take administrative and legislative steps for the enforcement of the fundamental rights of the people in the Northern Areas, and allow them to be governed by their elected representatives within a period of six months. The Supreme Court, however, declined to give a ruling on the request of the petitioners that the people of the Northern Areas should be given representation in the country's parliament. It observed that it could not decide on the form of government that should be set up in the Northern Areas to ensure compliance with the mandate of the Constitution. Following this, the Nawaz Sharif Government, then in power, announced a 'package' for the Northern Areas, which provided for an appellate court comprising three members of the Supreme Court bench to sit in Gilgit, and increased the number of seats in the council, which was renamed as the Northern Areas Legislative Council. The first elections to the NALC were held after the military takeover on 3 November 1999 and the second elections took place in December 2004.

Earlier, on 26 November 2000, the military government announced the delegation of financial and administrative powers to the NALC and increased the annual budgetary allocation for the area. In October 2002, the government decided to create two separate divisions in the Federal Government for dealing with the affairs of PoK and the Northern Areas, instead of the same division dealing with both as was the case until then. It was also laid down that while the Federal Government would continue to deal with Home Affairs, Law and Prisons in the Northern Areas, all other matters would be dealt with by the Northern Areas Legislative Council. The Deputy Chief Executive was given powers relating to the transfers and postings of government servants up to a certain level and it was decided to treat the Chief Secretary of the Northern Areas on a par with the Chief Secretaries of other provinces of Pakistan.

Thus, the position regarding the Northern Areas is as follows. First, the Northern Areas are de facto an integral part of Pakistan. The Federal Government of Pakistan retains the following rights;

- to collect taxes;
- to enforce law and order;
- to station the army in the territory;
- to make recruitment to the armed forces from the territory;
- to prescribe the school curriculum; and
- to resettle outsiders, including ex-servicemen, in the area, in order to change its demographic composition etc.

The Northern Areas are de jure not an integral part of Pakistan. Hence, the people of the territory have no rights against the state of Pakistan such as:

- to vote in the Federal elections and to be represented in the Federal Parliament;
- to control their budget through their elected representatives;
- to self-determination etc.

Moreover, the Northern Areas are still governed by the Frontier Crime Regulations (FCRs). These were first promulgated by the British in the FATA before 1947 and were extended by Islamabad to the Northern Areas after they were occupied in 1948. The FCRs are similar to the Criminal Tribes Act promulgated by the British in India, under which members of tribes notified under this Act had to obtain the prior permission of the police before travelling from their place of residence to another place and were required to keep the police informed of their movements. One of the first Acts of the Government of India after it became independent was to abrogate this Act, but in Pakistan, the FCRs continue even today. People of the FATA and the Northern Areas violating them are liable to punishments such as a fine, forfeiture of property, cancellation of gun license etc.

The lack of economic development

Though the Government of Pervez Musharraf claims to have increased the annual budgetary allocations for the Northern Areas from Pakistani Rs. 600 million to one billion, the Northern Areas continue to be in a state of neglect, with three-quarters of the population estimated to be below the poverty line. It provides water for Punjab's agriculture, but there are no adequate irrigation facilities for its own agriculture. There is always an acute shortage of electricity and only now has China been helping in the construction of some small power stations. About 70 Chinese engineers have been working on the construction of these projects. For want of power, there is no industry. Apart from the Karakoram Highway (1,284 km) constructed by the Chinese in return for Pakistan's transfer of some of the territory in the Shaksgam area of the Northern Areas to Xinjiang, no new road construction has been undertaken due to the difficulty of the engineering works involved, particularly for the construction of tunnels. The North Korean Army offered assistance for the construction of tunnels and in 2001 a North Korean team of military engineers visited the area for studying the feasibility. In the absence of any development, the principal means of livelihood for the local people was service in the armed forces and in the tourism industry. The recruitment to the armed forces from amongst the Shias of the Northern Areas has been cut down. The tourism industry has been affected post-9/11 by the fighting in Afghanistan and the attack on a tourist bus carrying European tourists by suspected Al Qaeda elements, while it was going to Xinjiang along the Karakoram Highway. The rigorous enforcement of the FCRs by the army has even closed the door for migration abroad for the people of the Northern Areas.

Movement for the right to self-determination

As a result of the factors mentioned above, a movement for the grant of the right of self-determination for the people of the Northern Areas has been steadily growing. It first made its appearance in 1994 and has slowly picked up momentum since then, helped by the anger of the Shias over the suppression of their rights and of the people as a whole because of their status as second-class citizens. The anger of the local population over the failure of the military leadership to adequately compensate the families of those who had enrolled themselves in the Light Infantry Regiments (LIRs) and who were killed during the Kargil conflict of 1999 has aggravated the feelings of alienation. Since Musharraf had classified those who occupied the Kargil heights as Kashmiri *Mujahideen* and not regulars of the Pakistan Army, he was unable to openly acknowledge their deaths and compensate their families.

A number of new political formations have come up demanding either an independent Jammu and Kashmir, with the Northern Areas forming part of it, or the grant of independence to the Northern Areas. Prominent amongst them are the Jammu and Kashmir People's National Party, which is active in PoK as well as the Northern Areas, and the Balawaristan National Front (BNF), which is active only in the Northern Areas, the Gilgit Baltistan United Action Forum for Self Rule, and the Gilgit-Baltistan National Alliance (GBNA). These organizations have been confining their activities to taking up their cause with the UN Human Rights Commission and other UN organizations, creating an awareness of their cause in the international community by bringing instances of violation of the human rights of the people of the Northern Areas to the notice of international human rights organizations, etc. Their attempts to wage a political struggle through the NALC have been handicapped by the ban imposed by the Pakistani authorities on anyone contesting the elections unless he or she signs a pledge that the Northern Areas is an integral part of Pakistan, which they are not prepared to do.

The BNF welcomed Musharraf's proposals for a solution to the Kashmir issue with India by treating the state as consisting of seven components, with the Northern Areas being one of them, and by reaching an agreement with India for the withdrawal of armed forces from them to be followed by an exercise to determine the wishes of their people on their future. According to the BNF this was the first time someone in the Pakistani military leadership had admitted that the future of the Northern Areas was still to be determined, and proposed the withdrawal of the Pakistani troops from there before determining its future. But Musharraf went back on it following a furore in Pakistan against it. Even many senior army officers were reported to have opposed his idea on the ground that this would affect Pakistan's national security.

Since 1948, no government in India has had a lucidly planned and openly articulated strategy to regain control of this territory. In fact, there is not even adequate knowledge on the Northern Areas in the policy-making circles of the Government of India and in the public as a whole. As a result, the Government's pronouncements relating to PoK, have come to be seen as referring only to the area that Pakistan calls Azad Kashmir. The Northern Areas stand in danger of disappearing from the consciousness of the international community. In a cover page article on the Northern Areas carried in 1988, the *Herald*, the prestigious monthly journal of the *Dawn* group of publications of Karachi, had described the Northern Areas as 'the world's last colony'. Twenty years later, it remains so.

8 Jammu and Kashmir

Contours and challenges of cross-border terrorism

M. M. Khajooria

Introduction

Terrorism is essentially a minority phenomenon. A noisy and fanatic minority seeks to impose its will on the majority through the medium of terror.

The threat or use of force and violent means for attainment of political objectives, such actions being intended to influence the attitude or behavioural patterns of a large group of people other than the immediate victims, would be an acceptable working definition of terrorism. Modern terrorism poses complex, non-conventional, fast changing and multifaceted challenges. The terrorists wage a war without borders that admits no rules, conventions or moral code, thereby placing the targeted legitimate state in a severe moral and ethical dilemma. Terrorism can, therefore, only be successfully countered with pragmatism, flexibility and innovation.

Cross-border movements are especially inescapable in the case of terrorism promoted by a neighbouring state in an area where the indigenous collaborators do not control any 'safe area' in strategic depth populated by committed supporters on the home turf, or where the resentment against the patron country is so intense and widespread that the refuge and assistance for the terrorists could be taken for granted. The sources of inspiration and facilities for training, equipment and funding, material management and of command and control in such cases have thus to be established in the 'secure' territory belonging to or held by the promoter state.

Background

The history of cross-border terrorism in Jammu and Kashmir can be traced to the massive tribal invasion of 1947 after Mohammad Ali Jinnah failed to cajole or coerce Maharaja Hari Singh into submission. The preference of Sheikh Mohammad Abdullah, the undisputed leader of Kashmir and his party, the Jammu and Kashmir National Conference, for a secular India against a theocratic Pakistan was no secret. In fact, the people of Kashmir, under the leadership of Sheikh Mohammad Abdullah, had decisively rejected the pernicious 'Two Nation Theory' on which the very edifice of Pakistan was constructed. This was made clear to Jinnah publicly and in no uncertain terms during his visit to the valley in May 1944. Caught between the vacillations of the maharaja and defiance by the sheikh, he, in sheer desperation, authorized the tribal invasion of Kashmir in 1947 to be planned and supervised by Pakistan military regulars. The Pashtun tribes were motivated to mount the cross-border assault in the name of 'Islam' and to save the Muslims of Kashmir from the atrocities of an infidel ruler, and for this purpose the services of mullahs and greedy and pliable tribal chiefs, who wielded substantial influence upon the illiterate and ignorant tribesman, were requisitioned. Ostensibly, they wanted to liberate their Kashmiri Muslim brethren but their (tribesmans') primary objective was rape and loot. The real face of *Jihad* was unmasked by Father Shank and nuns of Baramulla Mission Hospital to the world in the following words:

> The tribesmen, great, wild black beasts they were – came shooting their way down from the hills on both sides of the town. A twenty-year old Indian nurse, Phillomena tried to protect a Muslim patient, whose baby had just been born. She was shot dead first. The patient was next. Mother Superior, Aldetrude rushed into the ward, knelt over Phillomena and was at once attacked. The Assistant Mother, Teresalina, saw a tribesman pointing a rifle at Mother Aldetrude and jumped in front of her. A bullet went through Teresalina's head. At that moment Colonel Dykes, who had assured us that we would not be attacked, rushed from his room a few yards along the terrace to get the Mother Superior out of danger, shouting at the tribesmen as he ran. But the Mother Superior fell shot and Colonel Dykes collapsed beside her, with a bullet in his stomach. Mrs Dykes ran from her husband's room to help him. She too was shot dead. While this went on, Mrs. Geo Boretto, an Anglo-Indian, was killed in the garden before nine nuns. Then the nuns were lined up before a firing squad.

This was how the 'indigenous freedom struggle in Kashmir' began. It was another matter that Pakistan could not muster even a symbolic participation by the Kashmiris in this so-called *Jihad*.

The landing of the Indian army at the Srinagar makeshift airport after the maharaja signed the Instrument of Accession signalled the beginning of the end of Pakistan's first phase of the cross-border terrorism. The infiltrators represented a creed abhorrent to the Kashmiri secular identity and liberal humanist values integral to Sufi Islam, the religion that they practised. The call for *Jihad* by Pakistan was responded to by resistance and rejection. Fully supported by the people, and actively assisted by the National Conference cadres and local militia, the Indian army succeeded in repulsing the tribal invaders. They were soon on the run in full retreat: Pakistan had become a dirty word in Kashmir.

India moved the United Nations Security Council to broker a ceasefire agreement. Even though the guns were silenced on the borders the Pakistani rulers' obsession with Kashmir remained intact and undiminished. Pak military top brass were tasked to work out a politico-military strategy to carry forward the war behind the enemy lines 'in relentless pursuit of the national objective of making Kashmir an integral part of Pakistan'. This was said to have been settled even before the UN Security Council sponsored ceasefire agreement was formally signed on 27 July 1949. From day one, the responsibility for handling 'Kashmir operations' rested squarely with the Pakistani army.

The political crisis of 1953 that culminated in the dismissal and arrest of Sheikh Mohammad Abdullah and some of his colleagues was an answer to the prayers of Pakistan rulers who were quick to cash on this bonanza. The operatives from Pakistan began to make frequent forays across the Ceasefire Line. Contacts were established with dissident leaders of the National Conference and emotionally surcharged youth with a view to promoting subversion and sabotage. The Plebiscite Front cadres and youth were, for the first time, taken across the Ceasefire Line (CFL) by the Pak Field Intelligence Units (FIU) of the Pak army for indoctrination and training. It appeared that the second phase of the cross-border militancy in Jammu and Kashmir had commenced.

The deteriorating internal politico-administrative situation, developments in the international arena and the level of success by Pakistan agencies in promoting discontent, subversion and sabotage in the state, critically influenced secessionist activity and cross-border mischief in Jammu and Kashmir. It was minimal during the decade of development (1953–63) under the Prime Ministership of Bakshi Ghulam Mohammad. The liberal dispensation during the stewardship of G. M. Sadiq was misused by Pakistan's Inter Service Intelligence (ISI) to serve its ends. Misreading the people's mood, Pakistan resorted to sizeable armed cross-border infiltration as a prelude to the Indo-Pak war of 1965 in which Pakistan again did not succeed. The hostilities were halted after the Tashkent Accord. The next round in 1971, on the other hand, proved more costly for Pakistan. It not only suffered a humiliating defeat but also lost its eastern wing, which emerged as an independent country on the world map called Bangladesh.

The Kashmir issue lay on the back burner between 1972 and 1979. It was, however, retrieved and accorded the highest priority by Gen. Zia-ul-Haq who assumed complete control in Pakistan after the 'judicial hanging' of Z. A. Bhutto. 'The Mullah in Khaki' wasted no time in establishing a collaboration with the Pan-Islamic fundamentalist movement. Flush with petro-dollars and a vast reservoir of human resources they were more than willing to contribute to the *Jihad* in Kashmir.

The anti-India sentiment in Kashmir fuelled by political and administrative mismanagement was fully exploited by the 'dirty tricks' department of the ISI. Politically confused, callously deprived of avenues of higher education and job opportunities for ulterior motives by their own leaders, and systematically subjected to indoctrination of a distorted version of the religion, a section of youth became vulnerable to the vicious Pak game plan. The products of the fundamentalist *Jamaat-e-Islami* run *madrassas* provided the fanatically committed nucleus and ideological vanguard. ISI struck at the 'right moment'.

The sponsoring of the *Jihad* in Afghanistan and emergence of Pakistan as the key player in the unleashing of cross-border fundamentalist terror in that country and the decision of the United States to oust the former Soviet Union provided the opportunity, the wherewithal and strategic environment for duplicating the script and drama in Kashmir. The commencement of a proxy war in Jammu and Kashmir on 31 July 1988 with two explosions in Srinagar that coincided with the withdrawal of the Soviet Union from Afghanistan was no coincidence. By then, training camps had been established all over occupied Kashmir and parts of Pakistan for moulding the raw recruits into armed terrorist (*Mujahideen*). Facilities for advanced and leadership training courses and battle indoctrination were made available in Afghanistan to the *Mujahideen* for the Kashmir front. The internal situation was appalling. There were reports of conductors inviting passengers to board buses bound for terrorist training camps in Chakoti in Pak-Occupied Kashmir (PoK) by shouting *Chakoti Hey* in the Lal Chowk bus stand in Srinagar. It may, therefore, be seen that the common perception that terrorism suddenly 'erupted' in Kashmir in 1988 was flawed and contrary to facts. The general impression that the 'blatantly rigged' elections in 1987 were a product of this militancy, also represented only a small portion of the truth.

Contours

Cross-border terrorism in Jammu and Kashmir had two facets, external and internal. At the external level, the promoter country, the Islamic Republic of Pakistan, was engaged in diplomatic manoeuvres to build up world public opinion against the settled fact of the state's accession with India and to arrange men and material for the *Jihad* in Kashmir. Its special focus was upon the Western powers, especially the US. It kept on peddling the outdated and redundant UN resolutions and harping on the so-called 'right of self-determination' of the people of Jammu and Kashmir. Highly priced US lobbying firms were hired to supplement the efforts of front organizations of Kashmiri emigrees in the US, the UK and elsewhere. These outfits were designed, directed and financed by ISI. They promoted the cross-border *Jihad* as the 'struggle for freedom' by the 'oppressed Kashmiris'. Islamic countries also figured prominently on the Pakistani shopping list for goodwill and support of the Kashmir issue.

Pakistan heightened tension on the borders through regular bouts of artillery firing. This not only facilitated the infiltration by armed *Mujahideen* but also enabled Kashmir to be projected as the nuclear flash-point in South Asia. The object was to sustain and enhance international concern in the Kashmir dispute and secure third party intervention in a matter that both countries had undertaken to resolve bilaterally under the Shimla Agreement. Pakistan also provided trained manpower, training facilities in PoK, Pakistan and Afghanistan. The killing of over two dozen Kashmiri militants under training in the Khost training facility in Afghanistan during an American missile attack on a bin Laden hideout in the Clinton era conclusively established this fact.

On the internal front, the ISI–*Jamaat-e-Islami* axis strove to preach and enforce the fundamentalist and obscurantist interpretation of Islam and the subversion of the centuries-old composite liberal value system. The people were sought to be terrorized into submission through the most barbaric, brutal and horrifying killings. Political leaders, Islamic scholars and intellectuals were slain with the aim of breaking popular resistance. In short, the cleansing of ethnic and religious minorities, murders of opposition political leadership and communalization of the civil society were high on their agenda. The local terrorist agents and their political counterparts fomented and accelerated alienation against India, infiltrated into the state apparatus, pontificated against democratic and secular institutions and did everything possible to bring the Government of India into disrepute.

The external and internal manifestations of cross-border terrorism sustained and reinforced each other. Success or failure on both fronts determined the measure of the strength or debility of the overall campaign. Obviously, the politico-military situation within the frontiers of Jammu and Kashmir primarily dictate and grade the rise or fall in the graph of cross-border terrorism. In essence it is the cross-border dimension that provides oxygen for terrorism and keeps it going. It should also be recognized that the terror organized, sustained and enhanced from across the border/LoC was designed to ensure that the people of the state were unable to objectively, dispassionately and fearlessly even consider, much less exercise, their political options.

The Indian response

Both conceptually and strategically, the Indian response to the proxy war unleashed by Pakistan was flawed from day one. The no-holds-barred conflict launched by Pakistan over eighteen years before that admitted of no politico-military objectives and broad counter-thrust. The exercise of the military options, the designing and execution of political initiatives as well as the moves in the field of psychological warfare were inevitably coloured by this unrealistic and jaundiced appreciation.

The real aims of the enemy escaped the calculations of the Indian establishment or were glossed over, resulting in faulty or disoriented political, diplomatic and military response. They pursued the disastrous policy of chasing the events. The tardy and piecemeal response made the task of the enemy and its agents easy. The Government of India thus committed the mistake of doing too little too late. Pakistan took and continued to take full advantage of the initiative that had thus been dropped in its lap. The rest was history.

It should have been obvious that the terrorists aimed at terrorizing the entire population in pursuit of their politico-religious objectives through military supplies exported from across the borders. It was an undeclared war – a battle for the minds of the people. The state, however erroneously, treated it as a mere law and order problem to be tackled by the police, the army and the paramilitary forces. The absence of any official effort to mobilize public participation in the proxy war was a clear pointer to the lack of a sufficient comprehension of the threat, the contours of the enemy focus and the very dynamics of terrorism itself. All efforts to wake up the governments at Srinagar/Jammu and Delhi to their responsibilities on this behalf fell on deaf ears. One wonders how any informed creature could conceive of winning a battle for the minds of the people without creating awareness amongst them and obtaining their conscious and active participation.

The ease with which thousands of recruits, trained and well-equipped *Mujahideen*, managed to cross the Line of Control/International Border both ways along with hundreds of tons of ammunition and huge quantities of sophisticated communication equipment, placed a question mark on the capability and integrity of a segment of Indian security forces. The argument that it was impossible to 'seal' the tortuous and long borders in Kashmir came to be scandalously named as the 'terrorists' National Highway'.

Severe constraints of topography and the length of the borders notwithstanding' innovative, more effective and unconventional measures, results which could be quantified and monitored, should have been devised in the early days of the current phase of secessionist terrorism. The harsh popular verdict was that the security agencies had failed to stem this tide. The issue had, however, been pushed under the carpet of secrecy in the name of national interest. This would not do. There can and will be no holy cows in the realm of national security.

Landmarks

'Guest Mujahideen' in Kashmir

Cross-border terrorism assumed wider international dimensions after the Kashmir *Jihad* ceased to be Pakistan specific and was formally adopted by the Pan-Islamic fundamentalists thanks to the exertions of Zia-ul-Haq. From then on it not only became integral to the grand design to enforce *Nizam-e-Mustafa* world wide but also the instrument of Pakistan's state policy. Speaking at a conference in Delhi on 'Peace Dividend – Progress for India and South Asia' on 13 December 2003, former Pakistan Prime Minister Benazir Bhutto admitted that Islamabad had backed a low intensity conflict in Jammu and Kashmir during her first tenure as head of the Government of Pakistan in the 1980s. This, she claimed, was done to 'hype' the Kashmir issue 'and was the joint decision of the military and political leadership'.[1] In fact, the induction of highly trained, well-equipped and battle-hardened *Mujahideen* from Afghanistan into Kashmir at the end of the Afghan war was part of the integrated plans of *Jihad* encompassing both theatres. The fortnightly newspaper published from Jammu, which was the first to expose the Pakistani move to introduce foreign *Mujahideen* in Kashmir, reported:[2]

> About ten thousand fundamentalist Arab *Mujahideen* on the loose, after the change of government in Kabul are, it is reliably learnt being screened by the notorious Pakistan Inter-Services Intelligence with a view to select crack groups of Arab terrorists to infiltrate into Kashmir. The Arab volunteers have been activity fighting in Afghanistan along with Gulbuddin Hikmatyar's *Hizb-i-Islami* since 1978. About five hundred of them have so far been killed during the operations in Afghanistan. The United States' agencies have also impressed upon the Arab Government the need to halt forthwith all assistance to the fundamentalist elements amongst the *Mujahideen* operating in Afghanistan.

The foreign militants were first located in the Sopore area of the valley in June 1992 and interviewed by an Indian TV channel. They were two Sudanese brothers, both engineers by profession and experts in manufacturing improved explosive devices. They were followed by *Mujahideen* from other countries like Chechnya, Egypt, Algeria, Saudi Arabia and more importantly from Pakistan itself.

The *Shivalik Chronicle* also disclosed that:[3]

> General Hamid Gul, former Director of Pak ISI, who was forced to seek early retirement has boasted that 'no power on earth can stop Afghan *Mujahideen*' from going to the rescue of their Kashmiri 'brethren who were engaged in a grim battle for independence from Indian Yoke'. 'If the Afghans could throw out a super power,' he said, 'there is no reason why Kashmiris should not succeed in evicting India from Kashmir, which after all is merely a dominant power in the subcontinent.'

The diversion of foreign *Mujahideen* from Afghanistan to Kashmir enabled Pakistan to boost its waning militancy. It also averted the social and economic disaster that was bound to accompany the influx of thousands of battle-hardened fundamentalist fighters rendered jobless into the already beleaguered North West Frontier Province.

This arrival of 'Guest militants' changed the complexion of the terrorist campaign and led to the upgrading in military training within Kashmir. The alien mercenaries provided cutting edge leadership to militant activities in Jammu and Kashmir. New routes for infiltration, hitherto considered insurmountable, were opened by these experienced fighters. They enacted kidnappings of Western nationals to garner international publicity and recognition. Extremely brutal massacres became common. Civilians were skinned alive in Doda in public by Abdullah, the butcher, and entire hamlets torched with innocent men, women and children roasted alive for the sin of being 'infidels'. His counterpart, the Sudanese Abu Zubair, similarly shamed humanity by decapitating Muslim women and children related to men serving with security forces as non-combatants or simply and arbitrarily branded as informers in the border district of Poonch. He would kick the severed heads around like footballs with glee. Demographic change by terrorizing the remaining Sikhs from Kashmir and Hindus from Muslim-majority districts of Doda, Poonch and Rajouri in Jammu region to migrate was accorded high priority. Murderous assaults on Shia leaders followed as a logical corollary. By 1994, the foreign terrorists had taken over the command and control of the terrorist movement in Jammu and Kashmir.

The Kargil conflict

A lot has been spoken and written about the most significant and blatant cross-border thrust by Pakistani regulars and Pan-Islamist terror groups in the Kargil sector in June–July 1999. By the time (18 July 1999) the intruders withdrew, 410 Indian officers and Jawans were reported to have been martyred and 594 were reported wounded. The enemy suffered 689 casualties including 41 officers. The nation has since paid glowing tributes to the martyrs, expressed solidarity with the bereaved families and the then ruling party harvested electoral gains. Confusion, however, persisted about the hows and whys of the bloody and costly conflict. The Kargil Committee submitted a wishy-washy report which concealed more than it revealed. The responsibility for failing to anticipate and pre-empt the enemy plans was yet to be apportioned. That Pakistan was up to mischief through the Northern Areas was no secret. General Musharraf opted for the established pattern of a mix of terrorists and Pakistan army, the latter not only in the lead role but also performing under the normal chain of command and utilizing regular supply and communication lines.

That a Ladakhi shepherd had to wake up the authorities to the large-scale penetration of the LoC speaks volumes regarding the level of intelligence and vigilance. Only in India could a Defence Minister get away with an irresponsible and arrogant retort like 'You cannot expect us to have been in their operation room, to know exactly what their moves were.'[4] The Defence Minister forgot that the nation had a right to know exactly who bungled, what the precise nature of the blunder was and what punitive and corrective measures were taken. For this purpose a fresh, comprehensive and in-depth probe was called for.

Pakistan aimed at achieving three main objectives through its Kargil operation. First, to revive the sagging morale of the terrorists and diminishing terrorist capability due to the change of mood of the common man in Kashmir after the installation of an elected government (1996). On my return from a visit to Kashmir in July 1997, I had written in a daily newspaper published from Jammu:[5]

> There were times when terrorism reached out its tentacles like some grotesque flesh eating plant that sucked in the Kashmiri youth. This was no more the case. Today, the volunteers were almost non-existent and

parents got sufficiently frightened to 'donate' their sons for the cause of the phony *Jihad*. In fact, there were reports of ISI offering high cash incentives for recruitment. There was little doubt that the frontiers of terrorism in Kashmir had considerably shrunk.

If at all Pakistan gained some advantage in the sphere, it was insignificant and fleeting. The state Assembly elections of 2002 clinched the issue beyond any doubt.

The second objective was to gain territory in that difficult rugged terrain, and control of strategic heights with a view to not only interdict the Stringer–Leh highway but also to dominate Shia Kargil (16,036 km^2) that had remained impervious to the call for *Jihad*; also to open up access to the already terrorist-infested sprawling mountainous district of Doda (11,691 km^2). This would provide a link-up with Uttar Pradesh and Uttaranchal (via Himachal Pradesh) where the ISI network was strong. The defeat and withdrawal from the area on the Indian side of the LoC dashed this grandiose Pak dream to the ground.

Third, to internationalize 'the Kashmir Issue' so that the US could arbitrate in the 'resolution of the core issue of Kashmir'. It had to be conceded that Pakistan did achieve considerable success on this score. The Kashmir issue had been internationalized as never before. After meeting the Prime Minister and his deputy on 28 July 2002, the US Secretary of State, Colin Powell said in a press conference in New Delhi that 'Kashmir was on the international agenda'.

The Pakistani army regulars and terrorist elements indeed withdrew from the Indian side of the LoC only after President Clinton publicly exhorted Pakistan to respect its sanctity. That India had earned such a guarantee under the Shimla Agreement in exchange for 12,800 km^2 of Pakistani territory occupied by the victorious Indian army and 97,000 of her prisoners, became inconsequential. There was, however, no escape from the reality that since the Kargil conflict, the US had assumed the role of a facilitator whatever it meant, and was calling the shots. It was well known that India wanted the US to pressurize Musharraf to honour his pledge given through Colin Powell not to permit use of territory under Pakistan's control for cross-border terrorism. Pakistan, on the other hand, wanted the big brother to put pressure on India to sincerely and speedily address Pakistan's concern on the core issue of Kashmir to its satisfaction.

9/11 and beyond

The world community remained indifferent to the Indian complaint of Pakistan promoting cross-border terrorism in Jammu and Kashmir as part of international terrorist designs. The Western powers, especially the US and the UK, swallowed the patently false and mischievous Pakistan assertion that the trouble in the Indian state was indigenous and that Pakistan was merely providing diplomatic and moral support. The 9/11 (September 2001) terrorist attack on the US pride and world primacy changed everything.

In a way, the United States paid the price for turning a blind eye to the tremendous strides taken by Pan-Islamic cross-border terrorism when it was hurting others including the largest working democracy of the world. It chose to ignore the fact that Pan-Islamic fundamentalism was dedicated to the imposition of *Nizam-e-Mustafa* throughout the world. That the US was bound to come in conflict with this fanatical force sooner rather than later escaped the calculations of the policy-makers in that country.

When the 9/11 assault occured, President Bush responded quickly and firmly. He declared war on global terror and its bastion – Taliban-ruled Afghanistan – which sheltered the world's most wanted terrorist, Osama bin Laden. His postulation that those who were not with the US in its war against international terrorism were against it, put Pakistan on the spot. At that time, Pakistan was on the brink of anarchy. The sectarian violence promoted by the extremists was tearing apart the very fabric of their Islamic entity. The army itself was infected by the virus of fundamentalism. The American call provided Pakistan with both the challenge and the opportunity. Musharraf made the best of both and quickly jumped on the bandwagon of international anti-terrorist coalition.

His 'policy statement' on 12 January 2002 was superbly crafted. It managed to address a variety of concerns including those of India and world public opinion while firmly retaining the focus upon the interests of Pakistan. It also sounded like a call to his fellow countrymen to jettison the backbreaking and unsavoury baggage of fundamentalism acquired since the days of Zia-ul-Haq. An assurance to cry a permanent halt to the cross-border terrorism in Jammu and Kashmir was given by the Pakistan President to the US Deputy Secretary Armitage on 6 June 2002 in Islamabad. The international media and world leaders hailed this as a 'positive development'. Considering Pakistan's past record, this euphoria needed to be tempered with caution. It did, however, seem that the orders to halt cross-border movement were passed down the line through the ISI. A major general of ISI was reported to have conveyed these orders from Musharraf to 'militant commanders' from the front lines at an army base, 25 miles from the border.[6] One commander of militants was reported to have told *Newsweek* that they denounced President Musharraf by name. One of the commanders was reported to have shouted, 'After ditching the Taliban, Musharraf has now betrayed the Kashmir cause.[7] How can we accept this?' he demanded. The political Pan-Islamic parties, groups and individuals of eminence reacted with equal vehemence and anger. Strong and unequivocal condemnation of General Musharraf's commitment to permanently stop cross-border movement of terrorists and dismantle the terrorist training camps was voiced in the *Ijtimah* at Muzaffarabad (PoK). The congregation was attended by heavyweights like the former Chief of the Pakistani army, General Mirza Aslam Beg, former Chief of the ISI, Major General Hamid Gul and Qazi

Hussain, the Pakistani *Jammat-e-Islami* supremo. They strongly criticized the President for 'surrendering before Indo-American pressure' and vowed to catry on the *Jihad* in Kashmir. This rattled the Pak dictator who decided upon a tactical retreat.

General Musharraf flatly denied having given any 'assurance' to the United States to permanently end cross-border terrorism. In an interview to the *Newsweek*, he said, 'I have told President Bush, nothing is happening across the Line of Control. This is the assurance I have given,' and added, 'I am not going to give you an assurance that for years nothing will happen . . .' The US however, responded quickly with a categorical contradiction. The State Department spokesman, Richard Boucher, said on 25 June 2002 that:

> Deputy Secretary Armitage was given assurances by President Musharraf on 6 June 2002 that ending of infiltration across the Line of Control would be permanent. These assurances were also given to the Secretary of State, to the President of the United States in their conversations, as well as repeatedly to our representatives and our ambassador in Islamabad.

But did this make any difference?

Dr. Ayesha Siddiqi, the noted Pakistan columnist, aptly described Pakistan's relationship with the United States as 'one of dependency and subservience on the one hand and deceit on the other'. Pakistani nuclear physicist Parvez Hoodybhoy addressed the issue of Musharraf's duplicity more candidly. He wrote, 'To run with the hares and hunt with the hounds – and imagine that the world will not know – has become impossible.[8] Today, 'in spite of General Musharraf's soothing statements, there is little doubt that militant camps shelter under Pakistan's nuclear umbrella. Having operated openly for over a decade in full public view and with obvious state backing, only magic or a massive military action can eliminate them.' About the amenability of Pakistan's Kashmir policy, Hoodybhoy was forthright when he wrote:

> Whatever Pakistan might choose to think, 'the rest of the world remains incredulous of the continuing official Pakistan position that it provides 'only diplomatic and moral support' to the people of Kashmir. Earlier denials of military involvement in Kargil or of providing military support to the Taliban regime have hugely diminished Pakistan's international credibility.

And he significantly added:

> It is now a matter of survival for Pakistan to visibly demonstrate that it has severed all links with the militant groups it had formerly supported, to be firm about providing 'only diplomatic and moral support' and to implement what General Musharraf promised in his January speech.[9]

This sane advice predictably fell on deaf ears.

On his part, President Musharraf had cleverly delinked the 'Kashmir Issue' from his fight against Islamist extremists and sectarian fanatics at home – Pakistan's position on Kashmir remains unchanged. In his own words 'the moral, diplomatic and political support to the cause of Kashmiris struggle for self-determination will continue'. It should also by now be clear that the dominant section in the US establishment had tactically fallen in line with Pakistan's stance on cross-border terrorism in Jammu and Kashmir. No wonder Musharraf had the temerity to assert in London on 18 June 2003 that the self-styled *Mujahideen* indulging in brutal killings, rape and arson in Kashmir were in fact engaged in a 'freedom struggle'.

As of now, two distinct streams of cross-border terrorism were emanating from Pakistan soil, one controlled and directed by the ISI and the other operating under the flag of Pan-Islamic fundamentalist outfits like *Lashkar-e-Toiba* (LeT) and *Jaish-e-Mohammad*. While the Pakistan Government did not seem to have any control over the latter, the ISI maintained a close liaison with them. Both were, however, agreed that the terrorist pot in Kashmir must be kept boiling. When some particularly atrocious strike was executed in Jammu and Kashmir by the LeT, its front outfit *Al Mansoorian* or *Jaish-e-Mohammad*, the Pakistan establishment officially denounced the incident but nevertheless used it to highlight the urgency of addressing the core issue of Kashmir. Pakistan also considered cross-border terrorism as a vital lever for bargaining in any negotiations with India. Both were aware that the day cross-border movement halted or was even substantially minimized, terrorism in Jammu and Kashmir would collapse.

The Indo-Pak peace process has produced the blessing of ceasefire on the border in Jammu and Kashmir, which has mercifully held for over two years now. The Indian army did not retaliate to the sole violation in Poonch on 18 January 2005. Instead the Director General of Military Operations spoke to his Pakistani counterpart who promised to investigate. The Chief of the Army Staff himself played down the incident. While speaking to the media on 22 January 2005 at Jammu he called the incident 'small and of trivial nature'. These are positive signs and hold out the promise of peace on the International Border (IB)/Line of Control (LoC).

The ceasefire facilitated the completion of the ongoing fencing of the 750 km long IB/LoC. It was a significant and welcome addition to the multilayered anti-infiltration system. However, there is no scope for complacency. The 18 January 2005 intrusion by five *Lashkar-e-Toiba* ultras in Krishna Ghati, Poonch, who were gunned down well within the Indian area underscores the point. Interestingly, the recoveries included fence tampering equipment.

There is little, if any chance of Pakistan under Musharraf making a significant change in the current policy on Jammu and Kashmir. The coming days would witness continued engagement with India in the peace process with sharper focus on addressing the core issue of Kashmir simultaneously stoking the fires of cross-border terror. Pakistani moves on the diplomatic, political and military fronts must be accordingly anticipated and pre-empted.

The October 2004 visit of Pak scribes to Jammu and Kashmir under the aegis of the South Asia Free Media Association (SAFMA) for the first time brought the people of Pakistan face to face with the realities of the Kashmir situation. The dispatches and writings of prominent Pakistani journalists and the comments of a cross-section of people in Srinagar are eye openers. The Pak journalists hardly found any worthwhile sentiment in favour of Pakistan in Srinagar, the heart of the valley. The leader of the delegation, Imtiaz Ali, while talking to media persons in Srinagar on 8 October 2004 candidly admitted, 'We realized that the situation is not totally the same as is being projected by the state controlled media of India and Pakistan. The view-point of people in Jammu city is different from those living in Srinagar. Jammuites are closer to New Delhi while the Azadi sentiment is dominant in Srinagar.'[10] Upset with what he saw and heard in Srinagar, Nusrat Javed, special correspondent of *News International* demanded, 'If you people want Azadi, why should Pakistan invest men and money?' The dormant but dominant anti-Punjabi sentiment integral to the Kashmiri ethos surfaced when the JKLF Chief Yasin Malik sarcastically said, 'Punjabis on both sides are trying to decide our fate.' Mariana Babber, diplomatic editor of *The News*, questioned the representative character of the politicians and said, 'If Kashmiris want a place on the negotiating table they have to first decide who represents them.' Their subsequent writings and TV projections in the same vein created an upheaval in Pakistan. In essence, while India negotiates with the Pakistan Government, the real focus should be on the Pakistan civil society, which has begun to assert. The civil society alone has the capability of stopping recruitment of their boys for slaughter in Kashmir in support of elements who did not even want to join Pakistan.

Height and night dominance were acknowledged as the two vital determinants for the outcome of any military engagement on the ground. Given the terrain and the unconventional character of the conflict in Jammu and Kashmir, it has not been an easy task particularly when even the basic night-vision devices were not available to Indian officers and men. Now that the equipment is available, all efforts need to be made to snatch the advantage back for army.

The common man in Jammu and Kashmir yearns for peace. Even those who in the past sympathized with the 'cause' are totally disillusioned. The alienation with the central political establishment and dissatisfaction with the performance of the local administration, however, persists. A conscious policy of reconciliation and reconstruction reinforced by good governance offer the only viable solution. Sizeable financial allocation has been made by the Government of India for the development of the state. It needs to be ensured that the benefit of every penny spent on that account reaches the people. A contented and secure citizenry provides the best guarantee for denial of shelter to the intruders from across the borders. And to be victorious, India must be perceived by the common Kashmiri as the winning horse on whom he could safely lay a bet. Remember no one worships the setting sun.

9 Tibet and the security of the Indian Himalayan belt

P. Stobdan

Introduction

The prospects for the Tibetan cause remain uncertain as any hope for a free Tibet is becoming increasingly feeble, though the Dalai Lama continues to strive for a 'genuine autonomy' within China. Notwithstanding the worldwide support for the Tibetan spiritual leader, nothing tangible has moved ahead in almost the last five decades, and there is little hope that things will take a different turn in the near future. In fact, the celebrity status and the popularity of the 14th Dalai Lama have made the Tibet issue more and more apolitical. The attraction of Tibetan Buddhism has dissipated further the political drive needed to liberate Tibet.

The history and politics of Tibet have, over the decades, become obscure, elusive and mystical, something incompatible with contemporary politics. A major propaganda war has been raging over the nature of the Tibet–China relationship. Various interpretations reflect major discrepancies, and they remain highly politicized. China's present claims are based on the seventeen-point agreement signed after 1950 with the Dalai Lama, which recognized Chinese sovereignty over Tibet (contested by the Dalai Lama as being signed under duress).[1] The agreement, however, remains unchallenged by major nations of the world, which recognize Tibet as China's sovereign territory. The UN resolutions of 1959, 1961, 1965 and the 43rd Session of the UN Sub-commission on Human Rights 1991 therefore have had no significant effect. No significant resolutions were either passed or enforced by international action in the past.[2]

China, meanwhile, has consolidated its hold over Tibet through large-scale development campaigns. Efforts have been made to modernize Tibet and improve the socio-economic levels of 6 million Tibetans. This has been demonstrated by large-scale investments, subsidies, development of industry, agriculture, and education and training. Large farming and irrigation projects are being developed.[3] The Qinghai–Tibet railway, which started in 2006, is poised to alter the face of the Tibetan plateau. Besides, several other major transport projects have been completed. Education policy since 1980 has hastened Tibetan integration into the Chinese mainstream, and to mollify world opinion, China in 1987 recognized Tibetan as the official language, with even Chinese cadres being encouraged to learn it. However, university- and technical-level education is still imparted in Chinese. China's 'Western Development Campaign' seems to be further advancing this process. China's achievements in Tibet were highlighted in a *White Paper* on Tibet entitled 'Regional Ethnic Autonomy in Tibet' issued by China's State Council in May 2004.[4]

Meanwhile, in exile since 1959, the Dalai Lama and his people, have administered their own parallel state institutions and structures of governance based on democratic principles. This, they claim, is a necessary step to prepare for governing a future free Tibet. Broadly, the exile government in Dharamsala runs a parallel network of institutions to monitor and govern about 300,000 Tibetan refugees in India. These institutions are supposed to be run on democratic values, though the Dalai Lama has supreme power to decide on all policy matters.[5] His administration conducts its own foreign policy and diplomatic activities and has administration 'offices' in many world capitals. He travels regularly and meets world leaders to canvass support for his cause. The Tibet lobby is fairly strong and active in the US Congress and European Parliaments. It exerts considerable influence on the foreign policies of the Western countries vis-à-vis China, especially on the sanctions restricting Western arms sales to China. His administration also has contacts with the Islamic world and it maintains close associations with many ethnic, democratic and political movements in the world.

In exile the centrepiece of the Tibetan movement revolves around the preservation and promotion of the Tibetan culture. The Western world has shown great interest in Tibetan Buddhism, and as a result some 6,000 monks who escaped from China have Buddhist administration centres all over the world to propagate their scholastic traditions. Cristina Rocca, author of *Zen in Brazil* in her recent interview to the *Times of India* said that 'Westerners associate Buddhism with therapy . . . how to address suffering and anger in everyday life . . . associating Buddhism with environmentalism and non-violence.' Buddhism she thought 'goes well with Western individualism'.[6]

This paper evaluates and analyses the current status of the Tibet issue and the interests that are at stake for India, particularly the security ramifications of the prolonged stay of Tibetan refugees in India. Attempts have been made to draw certain scenarios evolving on the Tibetan front particularly in the post-Dalai Lama era. The conclusion offers some policy suggestions to monitor the activities of Tibetan refugees so as to forestall any possible security implications emanating from them.

Political overtures

China has made several political moves on the Tibet issue particularly in the post-Mao era. Beijing initially suggested the Dalai Lama to return and live in Beijing on condition that he accepts Tibet as an integral part of China. The

Dalai Lama's first talks with Beijing started in 1979 and he came close to accepting Chinese demands in 1988, when he decided to drop the demand for complete independence and settle for internal autonomy (Beijing retaining defence and foreign affairs) along the 'one-country, two-systems' approach. However, the hope got dampened following frequent riots in Lhasa from 1987 to 1989. Perhaps the Chinese assessment was wrong when huge crowds gathered to meet the Dalai's delegates visiting Tibet that confirmed his popularity. Behind the scenes contacts until 1995 did not bear fruit, possibly also due to the purges of liberal forces in China from 1989 onwards. Beijing rebuffed the Dalai Lama's September 1987 five-point peace plan and in June 1988 the 'Strasbourg Proposal'[7] that called for transforming Tibet into a zone of peace and protecting its environment. China repeatedly criticized the Dalai Lama as a separatist and traitor and his high-profile international visits were dubbed as a ploy to internationalize the issue and split the motherland.

Resumption of the Tibet–China dialogue

A new series of talks between the Dalai Lama's envoys and Chinese authorities started in September 2002. The talks, including the fifth round held in February 2006, have produced no concrete results. The Chinese Government, in fact, does not even openly acknowledge that a process of negotiation between the two exists. The Governor of Tibet stated in March 2006 that the 'talks had not yet resulted in substantive negotiations', but added that the door was open for more dialogue. He also said that 'talks cannot be called negotiations but simply dialogue, or contact'.[8] The Dalai Lama's interlocutors also revealed that talks were smooth, but 'major and fundamental differences' remain even in the approach.[9] The talks were probably of a confidence building nature, though it appears that Tibetans are negotiating only for 'genuine autonomy' even if it means for a limited area of Tibet. But Beijing still sees the Dalai Lama's move as a sham.

Nonetheless, it does appear that the talks, both open and clandestine, have sufficiently raised the level of mutual confidence and understanding. During 2005, there was evidently more optimism in Dharamsala when Samdhong Rinpoche, *Kalon Tripa* (Prime Minister of the Tibetan Government-in-exile) talked about having a 'specific agenda', 'definite time frame' and reaching a 'decisive stage'. It was, in fact, indicated that the Dalai Lama's administration would consult Nicholas Haysom and Yash Ghai of Hong Kong University to assist them in the negotiation.[10]

The Dalai's conciliation

Any hope towards a major compromise seems unlikely in the immediate future though there are compelling reasons on both sides to reach an early negotiated settlement. Three major developments that dramatically shaped the Dalai Lama's position in the recent years include:

- India's position on Tibet (Vajpayee's visit to China 2003) altered the nature and intensity of Tibetan resistance, and in particular deeply undermined the Dalai Lama's political strength;
- the Karmapa's emergence on the scene has upset the nature of the Tibetan movement; and
- the rise in Tibetan extremism since China's release of a *White Paper* on Tibet (2004) has hastened the Dalai Lama's effort for an early reconciliation.

The urgency on the Dalai Lama's part, therefore, is clearly discernible from some of his recent statements conveying his appreciation for developmental changes in China. Seemingly reconciled to Tibet's status as a province of China, he said that Tibet would get 'greater benefit' if it remained within the Communist nation as long as its distinct culture was preserved. In his speech on 10 March 2005, he said 'as long as I am responsible for Tibet, we remain committed to . . . not seeking independence . . . and are willing to remain within [the] People's Republic of China'.[11] Conceding that China was in a 'win-win' situation, he said 'we're not suggesting separation, [but] that Tibet becomes more prosperous within China – and that it is also in the interests of the people of China to preserve our cultural heritage'.[12]

On the anniversary of the 1959 annual National Uprising Day, the Dalai Lama went a step ahead while stating *inter alia*, his willingness to visit China on pilgrimage and to see for himself the changes and developments in China. Most interestingly he said:

> Tibetans – as one of the larger groups of China's 55 minority nationalities- are distinct in terms of their land, history, language, culture, religion, customs and traditions. This distinctiveness is not only clear to the world, but was also recognized by a number of senior Chinese leaders in the past. I have only one demand: self-rule and genuine autonomy for all Tibetans, i.e., the Tibetan nationality in its entirety. This demand is in keeping with the provisions of the Chinese constitution, which means it can be met. It is a legitimate, just and reasonable demand that reflects the aspirations of Tibetans, both in and outside Tibet. This demand is based on the logic of seeing the future as more important than the past; it is based on the ground realities of the present and the interests of the future.[13]

To allay China's suspicion further, the Dalai Lama's envoy to America, Tashi Wangdi, while speaking at a session of the Congressional Executive Commission on China on 10 March said that:

> The proposed visit by the Dalai Lama is a win-win situation and the Chinese leaders should have no fear as to what might happen if such a visit is allowed, because the Dalai Lama will do nothing except bring

> about better understanding, harmony and friendship. The visit would have a calming effect on the situation inside Tibet. We have no doubt the whole world will welcome such a move.

It was clear that the idea of the Dalai Lama's visit to China's Wutaishan temple in Shanxi province – a sacred mountain that enshrines Boddhisatva *Manjushri* (Lord of Wisdom) – was taken up by his envoys in the February talks. In an unprecedented move, the head of China's State Bureau of Religious Affairs, Ye Xiaowen announced on 4 April 2006 that 'as long as the Dalai Lama makes it clear that he has completely abandoned Tibetan "independence", it is not impossible for us to consider his visit. We can discuss it.'[14] It was a major turn-around and the boldest ever statement made on the Dalai Lama, that received wide coverage in the state-run *China Daily*.[15] However, the timing of the statement ahead of Hu Jintao's US visit created doubts about China's sincerity. Beijing's olive branch to the Dalai Lama also came against the backdrop of the growing Indo-US relations and their commitment for furthering democracy. Furthermore China's gestures included the release of a nun jailed for 15 years, and dropping charges against a US researcher.

From the Tibetan perspective, a possible deal with Hu Jintao, who remained Party Chief of Tibet from 1988 to 1992, could prove to be a blessing in disguise for Tibet. They hope China might strike a deal ahead of the Beijing Olympics in 2008. Their optimism is also based on changes within China, particularly greater freedom for religious practices. The revival of Buddhism within China and a growing number of Han Chinese turning to the Dalai Lama's teachings are viewed as signs of positive change.[16] A recent case in point is the holding of the World Buddhist Forum in April 2006 in Hangzhou and Zhoushan city of Zhejiang Province in China – the first such event since 1947 – with enthusiastic support from the Chinese Government.[17]

The Chinese, on the other hand, perhaps hope that the issue would vanish gradually due to the modernization and assimilation of Tibetans into Chinese culture, and the demise of the Dalai Lama would mollify the resistance movement. As such, China may be playing a waiting game, waiting for the Dalai Lama to age and die. However, as things stand, both sides are painfully aware that the power vacuum in the post-Dalai Lama era could encourage both schisms within the Tibetan community, as well as allow radicals to lead from the front. Besides, the possibility of finding a new mechanism for a resistance overseas also exists.

The Dalai's dissidents

The Dalai Lama faces many challenges within, as a section of his people are irked by how much he has conceded just to get a seat at the table. The Tibetan Youth Congress with 20,000 members remains strongly critical of his non-violent and 'Middle Path' approaches. It considers them to be a 'recipe for future crisis'. The Tibetan Youth Congress always warned that should Beijing not soon seize the opportunity for a solution during the Dalai Lama's life span, 'it would be [too] late to head off support for full independence'.[18] The Congress believes in armed struggle and has a strong support base in India and overseas. It remains critical of the Tibetan elite cutting deals with the Chinese. At the same time it does not oppose the Dalai Lama but distrusts the Chinese.

The current trend indicates the Tibetan Youth Congress radically changing its resistance method of protests from holding rallies and hunger strikes to more extremist actions. Though it is difficult to predict the future of Tibetan extremism, formation of radical groups is on the rise especially in the aftermath of 9/11. The Dalai Lama has so far strongly curbed the extremist expressions but signs of Tibetans resorting to violent methods are growing. For instance, in April 1998, a Tibetan Youth Congress activist, Thupten Ngodup set himself on fire on Janpath in New Delhi protesting against the failure of the Dalai Lama's peaceful approaches. Pictures of Ngodup in flames have motivated others. Tenzin Tsundue, a poet and a writer, scaled 14 floors of a five-star hotel in Mumbai in January 2002 when Prime Minister Zhu Rongji was inside. Again when Chinese Prime Minister Wen Jiabao visited Bangalore in April 2005, Tsundue climbed a tower at the Indian Institute of Science (IIS) where Wen Jiabao was meeting Indian scientists.[19] Although it was a political protest and not a terrorist one, he had managed to highlight the significance of the IIS – symbol of India's scientific advancement – as a potential terrorist target. It gave a cue to *Lashkar-e-Taiba* (LeT) to carry out a terrorist attack inside the IIS, killing one scientist and injuring five others, on 28 December 2005.[20]

Like Tsundue, numerous Tibetan youths are flocking to India from across the Himalayas to fight for freedom. The most vehement critics for autonomy are between 20 and 30 years of age. Pankaj Mishra of the *New York Times* wrote a disturbing commentary in 2004 about disheartened Tibetan youths talking of 'many Chinese embassies in the world that could be targets, naming possible sites with disturbing precision'. Mishra quoted a Tibetan who said, 'Look at Palestine and Israel. Such small places compared to Tibet, but the world pays them so much attention because of the *Intifada*, suicide bombers and Osama bin Laden. What has non-violence achieved for the Tibetan cause, apart from some converts to Buddhism in the West?'[21]

At least 150 extremists comprising Europeans, Americans, Latin-Americans and Tibetans are supposed to be committed to guerrilla activity inside Tibet.[22] They are said to be in Europe, such as the French-based *Vercors Tibet Resistances* which provides arms training to Tibetans. Tibetans are also reported to be receiving training in Switzerland, Taiwan and other places. Tibetan immigration to the US has increased in the recent years. A close coordination seems to exist between Tibetans and Uighurs. Hundreds of Tibetan Muslims[23] are supposed to be working in Pakistan, particularly in international banking sectors. A trend is visible among the middle and younger level Tibetans who are leaving their families and renouncing normal life to join extremist groups. Similarly, many Tibetans are moving from South India towards the northern borders and Nepal remains the main operational base for radical Tibetans.

Many analysts observe that the majority of Tibetans are gradually losing faith in the Dalai Lama's soft approach. A section of influential Tibetans and their followers are turning towards Karmapa for leadership, though his political game plan is not understood completely as yet. But at the same time another section, especially Tibetan youth, doubts the efficacy of monks and philosophers controlling politics. They disapprove of Samdhong Rinpoche's leadership and his Gandhian political strategy of *satyagraha*. Rinpoche's rejection of ethno-nationalism and insistence for Tibetans and Chinese to coexist peacefully is viewed as politically naïve. For example, Jamyang Norbu, a noted analyst, charged Tibetan leaders with having 'an imperfect understanding of the politics of nation-states and the Darwinian reality of our modern world'. This was precisely how the Chinese accused the Lamaist regime in the past for failing to modernize Tibet as it was ruled by a few 'feudal' and 'reactionary' aristocrats who together with monks 'oppressed by a majority population of serfs and slaves, mostly by saddling their minds with ritual and superstition'. Chen Kuiyuan, a Chinese technocrat who served in Tibet, said in 1977 that 'when the Dalai ruled Tibet, there was not a single regular school; children of the working people had no right or opportunity to receive an education, and more than 90 per cent of the Tibetan people were illiterate'.[24]

However, the Dalai Lama's administration remains committed to the 'Middle Path Approach' and says that 'they cannot afford to jeopardize their efforts to creating [a] conducive atmosphere for negotiation in fear of criticism and opinions from a section of Tibetan organizations or individuals'.[25] A few recent remarks by Rinpoche are indicative of a breakthrough. He said, 'China is changing and there are very positive signs in its overtures towards Tibet. We expect [a] major breakthrough(s) in [the] near future.' Tibetans, in recent years, have toned down their anti-China rhetoric to ensure that the talks that have resumed after a long gap are not derailed, i.e., banning demonstrations protests during Hu Jintao's April 2006 tour of the US, closing down a few of the Dalai Lama's offices in Europe etc., among other important measures. Surprisingly, Rinpoche also stated that 'China's Communist rulers are not so much an enemy, as the rest of the world and, perhaps the majority of Tibetans believe they are'.[26] Rinpoche thought railways to Tibet would spur economic growth and help 'Tibetan products find markets in the rest of China'. However, it does not appear that Beijing is reciprocating in the same manner. On 11 May 2006 China referred disparagingly to the Dalai Lama as a 'splittist' when he was visiting the Latin American countries.[27] It appears that Beijing, while negotiating with the Vatican to withdraw its Embassy in Taiwan, wants a similar step from the Dalai Lama – to sever ties with Taipei and recognize Taiwan as an integral part of China.

Tibet and India

India's position since 1954 has been consistent in explicitly recognizing Tibet as an autonomous region of China and its implicit endeavour was to wait for a time when the Dalai Lama himself feels that he should return to Lhasa. India's stand on Tibet was based on the point Nehru described as, 'Neither the UK nor the USA, nor indeed any other power is particularly interested in Tibet or in its future. What they are interested in is embarrassing China.'[28] The Government of India reiterated in 2003 by accepting that 'the Tibet Autonomous Region is part of the territory of PRC'.[29] The change in India's position has been widely debated, though the government has clarified that it had consulted the Tibetan leadership prior to Vajpayee's China visit in 2003.

In the past, India's moral stand of providing political/spiritual sanctuary to the Dalai Lama and his people was considered by the world outside as master stroke diplomacy, as much as a well thought out strategic investment to counter China. India's legitimate role and monopoly on the Tibet question was accepted internationally. The fact was appreciated that many important Indian figures, including politicians in private, have taken a stand inconsistent with the stated government position. It is well known that the Dalai Lama, over the years, has managed to gain the sympathy of many parliamentarians, academics, activists and the public at large. He and his people through material and spiritual support, have privately persuaded a wide section of Indian politicians to galvanize Indian public support for them. The recent case in point is the visit by the Congress Member of Parliament, Rahul Gandhi, to Dharamsala to seek the blessings of the Tibetan spiritual leader.[30] This has created a state of ambiguity even in the minds of well-informed Indian circles.

India's ability to sustain an ambiguous Tibet policy particularly weakened after the end of the Cold War. A marked change in India's position on Tibet was visible from the late 1980s, with consequent dilution of the Tibet factor from India's strategic framework. Correspondingly, Indian public opinion too has gradually shifted in China's favour.[31] In contrast, China has effectively gained strategic interests in South Asia, a point fully articulated in Indian strategic discourses. The Ministry of External Affairs, which handles the Tibet issue and maintains a close communication with the Dalai Lama administration, appears to be playing no direct role in the current Dalai–China negotiation process, though Rinpoche did say recently that the 'issue can be better resolved' with India's involvement.[32] Ironically, the Chinese seem to be taking the assistance of Indian political figures such as Nirmala Deshpande and Sitaram Yechury as intermediaries to communicate with the Dalai Lama.

Tibetan perception of India

The Tibetans consider India as *gyagar phagspay-yul* (enlightened land), an expression of spiritual inheritance from India since the seventh century through the transmission of Buddhism. The Dalai Lama has described the Indo-Tibet relationship as between the *guru* and *chela*. This is more an expression of a diplomatic proviso on the Dalai Lama's part. At the popular level, Tibetans despise India and remain dismissive about what is Indian, much akin to the sentiments shared by neighbouring Nepalis, Bangladeshis and Sri Lankans.

At the higher level of political thinking, a section of Tibetans believe in a 'keeping India happy' approach. They include monks trained in Indian Sanskrit universities and those who strongly believe in Indian ideals. The present Prime Minister, Samdhong Rinpoche, belongs to this group. Others, who sway power politics, believe in engaging India in a constant power play vis-à-vis China or vice versa. This group, mainly led by the Dalai Lama and Lodhi Gyari, visualizes a perpetual conflict between India and China. They tend to seek Western support for sustaining the Sino-Indian estrangement. Lodhi Gyari has close association with the US establishments and currently heads the Dalai Lama's interlocution with Beijing. The third group exercises the middle approach of both engaging and blackmailing India as a strategy to deal with the problem. To put it simplistically, the trend of Tibetan elite opinion has moved in the direction of pro-India (1950–70), pro-USA (1980–2000), pro-Taiwan (2000 onwards) and pro-China (in process).[33]

The Dalai Lama has been saying different things to different people on issues impinging on India. A strong perception prevails in India that Tibet never stood up for India in the past. When Lhasa had the opportunity, it had even laid claims to Indian territories, including Tawang, when India was weak.[34] Privately, Tibetans believe that a large section of the Himalayan belt falls in the Tibetan area of influence.[35] India had criticized the Dalai Lama on several occasions for making unfavourable statements about the McMahon Line and matters concerning Sino-Indian borders. To cut it short, any consequential claims of an independent Tibet over any Indian territory would be a matter of concern to India. Once Tibet is independent, India would not be in a position to prevent it from getting closer to other powers. The situation in this respect would become worse should China and the Dalai Lama reach a complete reconciliation. Inevitably, Tibet could easily claim that Tawang belongs to it. The Dalai Lama's remark in 2003 that Tawang has traditionally been a part of Tibet has added a new dimension. Tawang being the birthplace of the sixth Dalai Lama already figures in fresh Chinese articulations concerning their claim over Arunachal Pradesh.[36]

Tibetan refugees in India

The assessment that the presence of the Dalai Lama and his people in India would be a provisional arrangement remains unfounded. After over four decades, the activities of Tibetan refugees have become intertwined with India's domestic as well as external concerns. Their prolonged stay and the spiritual influence of the Dalai Lama has made the Tibetan refugees well entrenched in the Indian system, and over the years, they have enlisted considerable support and sympathy among wide sections of Indian political, bureaucratic and corporate establishments. The track record of Tibetan refugees suggests that they have not only taken full advantage of India's lenient approach but have also been indulging in various unlawful activities, which, if not controlled, are fraught with dangerous consequences for India's security.

Roughly around 120,000 Tibetan refugees entered India along with the Dalai Lama in 1959.[37] Additionally, 3,500 Tibetans on average have been entering India annually since then. They are theoretically subject to rules and regulations applicable to foreign nationals, i.e.: the Registration of Foreigners Act, 1939; Foreigners Act, 1946; Passport (Entry into India) Act, 1920; etc. Tibetan refugees enter India illegally through Nepal with the connivance of the Tibetan Refugee Centre (TRC) and the UN High Commission for Refugees (UNHCR) office in Kathmandu. Nepal has always acknowledged its right against illegal entry, but the UNHCR called for regulations to allow people to use Nepal as a transit point. Since 2000, the number of Tibetans entering India via Nepal has risen significantly. Despite the introduction of Special Entry Permits (SEPs), Tibetans tend to contravene the guidelines introduced by India. It is estimated that there are about 100,000 Tibetan refugees in India for which no account has been taken.[38]

In early 2006, at the Dalai Lama's *Kalachakra Tantra* initiation (5–16 January 2006) in Amravati, Andhra Pradesh, the Dalai Lama's security department had registered 8,600 Tibetans who had illegally reached India's heartland. The actual number may have been more. Similarly, a parallel ceremony of the Karmapa additionally brought thousands from Tibet to Bodh Gaya in the same month. Such developments, which have been taking place routinely, are a gross violation of India's sovereignty. Accepted that none of them were terrorists or Chinese spies, they were allowed to violate the law of the land and this was a major breach of national security. One is tempted to ask whether it is possible to send 8,000 Indians to China in a horde? The Dalai Lama's security department suspected at least 12 Tibetans were Chinese agents. Interestingly, 110 Han Chinese also attended the Dalai Lama's *Kalachakra Tantra* in Amravati.[39]

Ironically, over the years, his administration has evolved its own mechanism to facilitate entry of fresh Tibetan refugees. A perfect *modus operandi* appears to be existing between the Dalai Lama's set up and the Chinese authorities for facilitating movements of people in and out of India perhaps without the knowledge of the Indian authorities. His set up also operates travel agencies to facilitate the movements of Tibetans in India. That the Tibetan refugees with dubious credentials have been obtaining Residential Certificates (RCs) – showing themselves as born in India – from The Foreign Regional Registration Office (FRRO) with the assistance of the administration is well known. Moreover, Chinese agents and security personnel are known to have infiltrated the Dalai Lama's set up.[40] They carry authorized documents obtained by dodging or bribing Indian authorities. It is also a practice that many Tibetan refugees, having lived in India, some even having served in the Indian armed forces, opt to return to Tibet.[41] Understandably, the Chinese use them for intelligence gathering. Several Tibetan refugees frequently and clandestinely visit Tibet and other parts of China bypassing Indian laws and regulations. This naturally has security implications.

Tibetan refugees are known in India for their indulgence in scandalous illegal activities. In the guise of religious activities they are involved in trans-border smuggling, money laundering, and the smuggling of wildlife products, rare artefacts, antiques, etc. In February 2006, police caught some of the biggest Tibetan wildlife poachers operating in India.[42] Thousands of skins from tiger, leopard, fox, cat, etc., are being smuggled out of India to Tibet. India's biggest poacher, Sansar Chand, was earning Rs. 60,000 for each tiger skin prior to his arrest sometime back. In 2005, Maneka Gandhi's remark 'throw all Tibetans out of India, each one of them is a poacher' highlighted the menace of the Tibetan refugees' involvement.[43]

Yet another aspect is the colossal commercial interests and enterprises built up by the Tibetan refugees, much of it without government clearance. Many of them have procured tribal certificates, gained government contract works, as well as employment in government services. They seem to have learnt the art of doing business in India without actually having an obligation to abide by the Indian laws. Tibetans have built vast infrastructures all over India. The Dalai Lama's own administration, including the Dalai Lama Charitable Trust, has acquired enormous assets of property and land worth hundreds of crores of rupees in India. Tibetan refugees enjoy a dominating presence in Arunachal Pradesh, Sikkim, West Bengal, Himachal Pradesh, Uttaranchal, and Jammu and Kashmir by acquiring tribal land through *benami* transaction. Ironically, the purpose of large-scale construction activities by Tibetan refugees is not clear if they are here as refugees. In fact, there are suspicions about the possibility of the Chinese supporting innumerable Tibetan construction programmes including the social and cultural infrastructures all along the Indian Himalayan belt. There is an element of outside support that Tibetan refugees receive from the vast sources of international links, mostly from the US and the Western support groups. Apparently, most of these transactions take place violating the foreign exchange regulations. Such funding also carries the potential of misuse as they are sought for the purpose of developing Himalayan environment, people and culture.

The Dalai Lama's proximity to the West has been a source of concern in India. He has been making controversial statements on Kashmir, Arunachal Pradesh, the Ram temple and issues of India's domestic concerns. He has also been accused of directly interfering in the affairs of the Indian Himalayan belt. Such utterances by him on critical issues like Kashmir appear to indicate that he is being guided by outside forces to embarrass India. For example in August 2001 the Dalai Lama created a controversy suggesting 'right to self-rule' for Kashmir while speaking at the South Asia Peace Conference in Chennai sharing a platform with the All Party Hurriyat Conference.[44] He had suggested that India and Pakistan should follow the 'middle path' similar to what he was doing with China, therefore equating it with the anti-Chinese struggle by Tibetans and the armed separatist drive in Kashmir. The Dalai Lama's stand that the people of Kashmir should be consulted while finding a solution to the Kashmir situation had pleased the Hurriyat leaders[45] but it sent shock waves amongst Indian political circles, both the then ruling Bharatiya Janata Party (BJP) and Congress condemning it as an act of direct interference in India's internal affairs.[46] The BJP termed the comments as 'incorrect' while opposition Congress said it was 'unwarranted'. 'He is not supposed to talk against China from Indian soil but instead he has chosen to speak against India,' a senior Congress leader said. The then Union Home Minister L. K. Advani also expressed concern and had ordered a review of[47] whether the Tibetan leader had violated the conditions of his asylum in India. He also made the Ministry of External Affairs remind the Dalai Lama that he was an 'honoured guest'.

The Dalai Lama and his people later tried to put an end to the controversy by reiterating that 'Jammu and Kashmir is an integral part of India'.[48] The Indian public also reacted as a mob and stoned the Dalai Lama's vehicle while he was travelling from Jammu to Dharamsala. This, in fact, created more dismay amongst the Dalai Lama's people. Yeshi Phensok, President of the National Democratic Party of Tibet, criticized the government's 'indifference and inaction' at the incident and said 'we are shocked to see that the Indian Government, which created so much fuss about the Dalai Lama's statement on Kashmir, is in no way bothered to even inquire into the circumstances that led to this incident in which he allegedly became the target of the Bajrang Dal's fury'.[49]

On 8 January 2004 the Dalai Lama created yet another controversy by accusing politicians of scoring mileage out of the Ayodhya dispute, and urged leaders to resolve it through dialogue.[50] When he expressed his willingness to initiate a dialogue, the BJP welcomed it but several Hindu leaders, including Kanchi Shankaracharya and Jayendra Saraswati questioned his propriety and said 'the Dalai Lama does not believe in Lord Rama. Nor does he belong to this land. How can he promote a dialogue between the two groups?'[51] The Muslim groups also opposed the Dalai Lama's interference. In an open letter, one Muslim wrote, 'Indian governments of different shades and people of India, in general, have shown immense respect and hospitality towards you and your people . . . I regret to write that developments of the last fortnight have cast an ominous shadow . . . that you have joined hands with the forces that want to undo the democratic secular India.'[52] Such utterances have not gone down well amongst the Indian public and in fact the Dalai Lama's *locus standi* in this respect has been strongly challenged.

Tibetanization of the Indian Himalayan belt

Despite the Indian Government's effort to rehabilitate the Tibetan refugees in the interior areas, especially in Central and South India, there has been a trend of Tibetan refugees shifting northwards in the Indian Himalayan belt. The Dalai Lama administration has admitted that Tibetan refugees are leaving their original rehabilitation camps for places where they have better conditions in which to live and practise their religion and culture. Over the years, this pattern has practically resulted in converting the Indian Himalayan belt into a zone of Tibetan influence where the Dalai Lama has established himself as an undisputed leader.[53]

The consolidation of Tibetan refugees in the Indian Himalayan belt indicates a pattern and feature, indicating subtle Chinese support for such a move. The Chinese, after having failed to suppress Lamaism, have started to use the institution to meet their end. Some Chinese leaders, including Hu Yaobang (1980), had candidly admitted the negative impact of the Cultural Revolution on Tibet.[54] Many of them realized that the quelling of Lamaism and emasculating the Dalai Lama had entailed rage and embarrassed China globally. Since the early 1990s the Chinese, therefore, opted for manipulating rather than stifling religion. In fact, Beijing had held this policy long in reserve while having engaged selected Tibetan Buddhist thinkers in the government-owned Chinese Buddhist Association. Previous Panchen Lama and Phuntsok Wangyal, who was a prominent Tibetan Communist, played that role.[55] The Dalai Lama remained worried about his chosen eleventh Panchen Lama being kidnapped by the Chinese authorities and replaced with their own candidate.[56]

The Chinese recognize that the Dalai Lama has ingeniously carved out a space for his leadership in the Indian Himalayan belt. Contrary to the consideration in India that the Dalai Lama's presence would help steer anti-China sentiments among people, the Chinese perceive that his activities in the Indian Himalayan belt are serving China's interests better.[57] There is merit in this point because as long as China controls Lhasa – the epicentre of Lamaist Buddhism – Tibetanization of the Indian Himalayan belt augurs well for China. For instance, in a scenario of the next Dalai Lama to be based in Lhasa, the people in the Indian Himalayan belt would look towards Tibet for allegiance. Inevitably, such a situation could become a destabilizing point for India's security interests.

Taking a cue from this, the Chinese have encouraged other influential Lamas, including Urgyen Thinley Dorje, the seventeenth Karmapa, to take shelter in the Indian Himalayan belt supposedly with their financial support. Urgyen Dorje's entry into India, in fact, has reinforced Tibetanizing a sizeable portion of India's eastern Himalayas, including Sikkim.[58] The Chinese have long been envisaging a Lamaist influence in the former Himalayan kingdom. They have carefully worked out a plot to have Urgyen Dorje enthroned in the Rumtek monastery with the connivance of their confidant

Tai Situ Rinpoche. Moreover, Urgyen Dorje's presence in Dharamsala has raised the frequency of visitors coming from Tibet, especially from the Amdo and Kham regions, on the pretext of meeting the Dalai Lama and Karmapa. They include top Tibetan religious officials carrying messages to the Dalai Lama administration, and in the recent years such interactions have intensified the trans-Himalayan networks.

Is there any solution for Tibet at the expense of India? To pursue their goal, Tibetans have intensified their activities since the 1980s in the Indian Himalayan belt, which is a compact geo-cultural region and a safe haven. Privately, Tibetans claim a list of territories extending from Bhutan to Ladakh as their area of influence. There is also the factor of the Tibetan resistance movement having religious rather than political overtones. A survey conducted in 1974–75 found that their anxiety over not being allowed to remain practising Buddhists in Tibet came top in importance.[59] This leads to the theory of Tibetan refugees achieving the objective without necessarily getting independence while creating a Tibetland in India. The following trends indicate this point:

- proliferation of Tibetan monasteries;
- illegal procurement of tribal land and properties;
- establishment of business and commercial assets;
- procurement of tribal certificates and contract licenses;
- entry into government jobs;
- mobilizing public opinion for Tibetan solidarity;
- organizing seminars and conferences;
- propagation of the Tibetan language;
- inclusion of the Tibetan language under Schedule VIII of the Indian constitution.

Over the decades, almost all significant Tibetan Lamas have found sanctuary in India. One of the most important trends is the construction of innumerable Tibetan religious buildings. The Dalai Lama administration claims that the Chinese destroyed over 6,000 monasteries in Tibet and it has been the task of the Department of Religion and Culture to replicate them in India. Over the years about 200 monasteries destroyed in Tibet have been rebuilt in India, Nepal and Bhutan.[60] (See the table overleaf.) In fact, the number appears to be much higher.[61] It is not clear why the Tibetans are engaged in such large-scale construction, spending millions of rupees, if they are here as refugees.

According to a survey conducted by the Integrated Development Plan (IDP) of the Government in Exile, 60 per cent of monks are those who joined after 1980 (seven per cent a year). The report says that over 5,000 monks entered India since 1980 thus forming a third of the monastic population in exile. The flight of nuns from Tibet has increased even more dramatically by over four times since 1980. It only suggests that their forays into India have a definite design.[62]

Table 9.1 Tibetan monasteries and nunneries

Sects	*No. of monasteries*	*No. of monks/nuns*
Nyingma	55	3,962
Sakya	20	1,135
Kagyu	36	2,056
Geluk	65	9,924
Bon	5	299
Total	181	17,376
Nunneries	8	549

Source: Tibetan Government in Exile official website http://tibet.com

Having found a footing in the Indian Himalayan belt, the Tibetan Lamas have taken control over almost all of the Indian monasteries from Ladakh to Arunachal Pradesh. This is a matter of serious concern in the entire Indian Himalayan belt. Another disturbing aspect is the introduction by them of sectarian differences along various fault lines such as pro versus anti-Shugden,[63] pro versus anti-Karmapa,[64] pro versus anti-Dalai Lama amongst others. As a result factionalism and resorts to violence have threatened the peace and stability in the Indian Himalayas. The prolonged stay of Tibetan refugees also has a negative impact on the socio-political fabric of Indian societies. The Dalai Lama's association with Westerners and of late with the Chinese, Taiwanese, and Cantonese amongst others suggests a concerted effort of an enhanced Tibetan influence in the Indian Himalayas.

The Dalai Lama administration supports several Indian Non-Governmental Organizations (NGOs) like the Himalayan Committee for Action on Tibet (HCAT), to keep the tempo of support and mobilize the Himalayan people in favour of Tibet and Tibetans. The HCAT strives for a pan-Tibetan movement and its demands, *inter alia*, to include the Tibetan (Bhoti) language under Schedule VIII of the Indian constitution.[65] The Dalai Lama also supports, in a subtle form, various intelligentsia in the Indian Himalayas for the propagation of the Tibetan language.

The destabilizing factor

Several studies have revealed eruptions of ethnic tensions in the Indian Himalayas between the local inhabitants and the Tibetan refugees. According to a study by Tanka B. Subba in the 1990s, tensions were higher in the Indian Himalayan belt than in South and Central India.[66] The Tibetan refugees' instability is more in Himachal Pradesh, Jammu and Kashmir, Arunachal Pradesh, West Bengal and Karnataka. In Himachal Pradesh it often became a law and order problem that even made the Dalai Lama think about moving from Dharamsala.[67] He had threatened to do so on 3 May 1994 after local residents attacked Tibetans in Dharamsala over the killing of a young Gaddi man by a Tibetan. Subba's study attributed this to the prosperity of the Tibetan refugees compared to the local inhabitants. This, according to others, is mainly due to foreign sponsorships and the Tibetan refugees' illegal means of earning. Subba noted that resentment stems from the procurement of land by Tibetan refugees on *benami* transactions in tribal areas.[68] In some areas they have acquired over 1,000 acres on lease and even started employing local tribal people as agricultural, domestic and business assistants. This has been resented by the local people who refer to the Tibetan refugees as 'exploiters'.[69] There have been wide protests from a number of organizations, which allege exploitation of local tribals by the Tibetan refugees. The phenomenon is becoming increasingly common in Karnataka.

Among other things, Tibetan chauvinism and their superior attitude over the people in the Indian Himalayas have also sharpened conflict in this frontier area. Subba's report suggested that there has been virtually no Tibetan refugee assimilation into Indian culture and social institutions. Moreover, concentration of Tibetan refugees along the border regions is getting complex. In fact, the intensity of tension and the Tibetan refugees as a destabilizing factor was higher in Bhutan. In 1979, the Tibetan refugees carried out a failed *coup d'etat* in Bhutan to overthrow the king and bring the Himalayan Kingdom under the control of Tibetan refugees. One of the objectives was to convert Bhutan into a military camp and a staging area for raids into China. The Bhutanese authorities arrested more than 30 Tibetan refugees in 1973 and in 1979, Bhutan's National Assembly passed a resolution under which the refugees who did not accept Bhutanese citizenship would be deported. The Bhutanese Government subsequently expelled 4,000 Tibetan refugees.

This Tibetan superior attitude has so far failed to give rise to unrest due to the respect and veneration among Indian authorities and politicians for the Dalai Lama, but the undercurrents of tension as elaborated above remain very strong. The situation in this regard could become explosive particularly in the post-Dalai era.

Future scenarios

The Tibet issue and Tibetan refugees in India could become highly explosive in the medium and long term and has the potential to become a major source of long-term contention between India and China. The activities of Tibetan refugees in India generally remain imperceptible. They are mostly unguided and risk dangerous consequences for India's security. Some plausible scenarios which could emerge are the following.

- As the prospects for a free Tibet recede, the majority of Tibetan refugees would opt for becoming permanently established in India and hence become India's exclusive responsibility. The Tibetan refugees have acquired vested interests and built colossal assets and infrastructure in

India. Their prolonged stay would have a destabilizing effect on the Himalayan region.

- The Dalai Lama is over 70 years old and he has not made up his mind about his next birth though he had indicated in the past that he might terminate the Dalai Lama's lineage.[70] China certainly treats him as a strategic commodity and will influence both the process of the next Dalai Lama selection and the role he would play in Tibet. They will seek to install their own nominee to neutralize the influence of Tibetans in exile. Since Lamaism is intertwined with politics, it is often the legitimacy rather than authenticity which determines the acceptance of the Dalai Lama by the people. In such a scenario, there will be two Dalai Lamas, one each in India and China. China then might decide that it will have nothing to do with the Dalai Lama born in India, and India will land up with a Tibet problem of its own which may have other implications.
- Chinese efforts to gain control over the Tibetan refugees would intensify in the post-Dalai Lama period. They have started manipulating Lamaist institutions to neutralize the Tibetan resistance. Many influential Lamas, including the Karma Kagyu sect – which is the most vulnerable to manipulation – have been successfully seduced or won over. Urgyen Dorje's emergence on the scene has changed the political, sectarian and regional realignments in the Tibetan refugee community. In the event of the Dalai Lama's demise, there could be much turbulence in India amongst the faction-ridden Tibetan refugees. The Chinese especially could create confusion and division along various fault lines. Both Shugden and Karmapa controversies have a potentially destabilizing impact in India.
- The undercurrent of resentment vis-à-vis the Government of India's decision to prevent Urgyen Dorje's control over Rumtek is growing in Sikkim. The Joint Action Committee of the All Sikkim Buddhist Organization (JAC-ASBO) among others has been attempting to get Urgyen Dorje to Rumtek. In the interim, the Chinese could, for instance, exploit the denial of Urgyen Dorje's claim to the Rumtek Monastery to entangle it with a section of population opposed to Sikkim's merger with India. Events associated with this could endanger the stability in the sensitive Indian Himalayan belt.
- The possibility of Tibetan radicals such as the Tibetan Youth Congress (TYC), Chushi Gangdrug (CSGT), GU SHU SUM (GSS), the Khampas and other groups resorting to violent struggle and engaging in cross-border operations cannot be ruled out in the post-Dalai Lama scenario. Extremists operating from India also cannot be overlooked. Although no Tibetan refugee would wish to harm Indian interests directly given their vulnerability to external influence compounded by feelings of 'let-down by India', such a situation arising is not impossible.[71] Should they resort to armed struggle, Beijing could build up a case that India is letting saboteurs and terrorists into Tibet. There are opinions that the current

events in Nepal could have possible implications for future developments in Tibet.

- The Dalai Lama wishes to seek a negotiated settlement with China during his lifetime. He has recently committed not to press for independence of Tibet provided he is left in charge of Tibetan affairs. He has expressed satisfaction on the renewed contacts with the Chinese and remains hopeful of finding a solution, as he sees China becoming 'more responsible and self-assured'. His desire to go on a pilgrimage to Wutaishan Temple in China and Beijing's positive response to that could possibly bring a substantive change in the stalemate. India needs to monitor the China–Tibet dialogue carefully.
- As the differences narrow down, China may be inclined to strike a deal ahead of the Beijing Olympics in 2008. A possible deal could lead the Dalai Lama to endorse the Chinese position on a number of issues impinging on India's security, including that of Arunachal Pradesh. Such a deal could also lead to most of the Tibetan refugees in India coming under Chinese control.[72]
- As the Sino-Indian border negotiation progresses, China may seek to dismantle the Dalai Lama's administration as a pre-condition for the final settlement. It is not inconceivable that China at some stage may pressurize India on the same lines as Nepal.
- As India opened the border-trade through Nathula some time ago, the possible influx of Tibetans into Sikkim can complicate India's position in the state. Various actors, including the Karmapa-Situ faction, are gearing up to exploit the situation and pursue their sectarian agenda.
- Linking of Lhasa with the Chinese railway has fundamentally altered the security dynamics along the Sino-Indian frontiers.

Conclusion

Given the sensitivity of the Tibet issue impinging on India's security and the rapid changes in the activities of Tibetan refugees, India should consider taking, *inter alia*, the following measures.

- Its present ambivalent Tibet policy may risk damaging relations with China and creating mistrust among Tibetans. The decision to retain the Tibet card needs to be decisive and clearly defined.
- While it needs to watch the emerging Chinese strategy closely, it also needs, as a matter of urgency, to contain the existing and potential adverse fall-out of the present liberal regime extended to the Tibetan refugees.
- It needs to evolve a refugee law to deal with the influx of people, including the Tibetans, entering India illegally. All refugees, including the Tibetan refugees, should be treated equally. The Government needs to tighten measures against illegal entry.

- As Nathula opens for cross-border trade, it must monitor and restrict the Tibetan refugees migrating for settling down in Sikkim.
- In view of the Dalai Lama's administration seeking a political support base in the Indian states, the government needs to carefully scrutinize and remain watchful of NGOs supporting the Tibetan cause in India.
- Influential Kagyue, Sakya and Nyingma Lamas are strengthening their hold in the Indian Himalayan belt. The growing Tibetan influence in the border areas may ultimately help China in the long run. The government needs to monitor the activities of the influential Tibetan Lamas carefully in sensitive areas. A number of Indian monasteries have already been taken over by the Lamas. The government needs to ascertain how many Indian monasteries are being controlled by them.
- The Dalai Lama has attempted many times to interfere in the internal affairs of India while making controversial statements. He is also seen directly interfering in the affairs of the Indian Himalayan belt and the UTD may also gradually start interfering in the same manner. They should be advised to desist from making anti-Indian utterances.
- It needs to be noted that the Karmapa (UTD) poses far more of a security challenge to Indian security than the Dalai Lama. While it must try to win over the young Karmapa, it should not give him undue importance, as he is unlikely to be accepted as a leader by the majority of Tibetan refugees.
- While it must encourage the Dalai Lama's reconciliation attempt with the Chinese, it should press him to take into account India's sensitivities on the border issue.
- It must start thinking about the post-Dalai Lama situation both from the point of view of domestic implications and the measures required to safeguard its interests in regards to the succession issue.
- Tibet will always remain vital to India's security interests and a moderating factor in its relations with China. However, improved relations with China will diminish the willingness to play India's psychological asset of the Dalai Lama vis-à-vis China. At the same time, India needs to recognize that the imperatives of geography, history and culture would keep Tibet inextricably bound with India. Asian Buddhist societies look towards India as their natural ally. Besides the Dalai Lama of Tibet, it currently hosts Jebtsundamba (god-king of Mongolia) and Druk Skyabgon Nawang Namgyal (god-king of Bhutan). India needs to view them as strong assets and use them skilfully in its diplomacy.
- It needs to encourage Tibetan studies, to maintain and support a second-track approach on Tibet.

In the light of the factors mentioned above, India needs to create a Tibet cell to coordinate policy matters with a long-term perspective. The US and others have appointed a coordinator on Tibetan affairs, not as a mark of support to the Tibetans, but to deal with the complexities of the issue, which at times are beyond their comprehension with fallout difficult to predict. The Tibet cell should be tasked to monitor holistic developments pertaining to Tibet and Tibetans in India and to provide inputs on what needs to be done from time to time.

10 The India–Nepal open border

Nature, issues and problems

B. C. Upreti

Introduction

An open international border is a rare phenomenon. There are very few countries in the world that have an open international border. What are the factors and forces which encourage countries to opt for an open border? Whether it is a source of enduring relationships or a cause of problems, tensions and conflicts is an important issue. The Indo-Nepal open border is a significant factor in the relations between the two countries. It has facilitated their peoples in various ways. However, there are problems and constraints related to an open border. It is for these reasons that it has the potential to turn into a source of tension between the two countries.

This paper seeks to focus on various positive and negative factors related to the Indo-Nepal open border. It also provides an analysis of the perspectives of the two countries on open borders and the possible ways and means to manage and regulate the border in the context of emerging issues and problems.

Evolution of the Indo-Nepal border

Before the advent of the British in South Asia and the establishment of their rule over India, there existed several big and small kingdoms in both India and Nepal. These kingdoms were shaped and reshaped through conquest and conspiracies. As a result there were shifts in state boundaries. Not much is known about the existence of border disputes between the countries surrounding India and Nepal. It is also a controversial issue whether Nepal ever had the status of a tributary state of India or that it ever became a part of the Indian empire at any time in history.[1] It can, however, be safely said that before the consolidation of British power in India and other parts of South Asia there existed a border between India and the nearby principalities of Nepal. It is possible that the shift in the border took place due to the annexation of territories on either side.[2] Understandably, there was full movement of people across the border even during those days.[3]

King Prithvinarayan Shah of the Gurkha principality founded the kingdom of Nepal in 1769, through a process of annexation and assimilation of the Baise and Chaubise states, Kathmandu valley and other smaller states of the then Nepal. The Gurkha rulers followed an expansionist policy after the consolidation of the kingdom of Nepal. They failed in the north but succeeded in the south and extended their rule in Kumaon and Garhwal and up to the Sutlej river in the west and Teesta river in the east. In the Anglo-Nepal war of 1814–16 and the subsequent Treaty of Sagauli (1816), Nepal lost the occupied territories. According to Article II of the Treaty, Nepal had to return Kumaon, Garhwal and the areas west of the Gandaki river and in Tarai between Rapti and Kosi.[4] Thus, the Treaty of Sagauli proved to be a turning point in the history of Nepal as it delineated and demarcated its southern boundary and fixed it between the Mechi and Mahakali rivers.[5]

As recognition of the Rana Prime Minister Jung Bahadur's assistance to the British against the 1857 mutiny in India, it was decided to give back western Tarai to Nepal. It was formally returned in the Treaty of November 1860. There was some dispute between the two countries over the boundary between Mondia-ghat and Banbasa in the Mahakali region which was eventually resolved in 1875.[6] Thus the settlement of the Indo-Nepal border was finally completed.[7]

Border demarcation

Jung Bahadur Rana was conscious of the need to demarcate and delineate the border with India and he made some effort in this direction. But the actual survey work of the border was undertaken in 1926–27 in the Geological Survey of India. It was the first time that aerial mapping of the border took place. Boundary pillars were erected at a definite distance all over the border to demarcate it. Where there were rivers flowing on the border line, the middle of the river formed the international boundary. It was decided to supervise and survey the international border at regular intervals so as to re-erect the missing pillars and resolve if there was any dispute over the border at the local level. For the purpose of the maintenance of boundary pillars the responsibility to maintain those of odd numbers was given to Nepal and those of even numbers to India. After the British withdrawal from the subcontinent, this arrangement with regard to the maintenance of the border between India and Nepal continued in principle. There were complaints and counter-complaints regarding displacement of boundary pillars, missing pillars, etc. Attempts were also made on several occasions to fix the displaced pillars in their original position. In 1981 the Indo-Nepal Boundary Commission was constituted. It was decided that the Commission would hold two meetings a year to look into various issues related to the border. There are no serious boundary disputes between India and Nepal except for the Kalapani region that was raised in Nepal during the UML regime in 1995. Nothing has happened on this issue since then, though it is still in the minds of several political groups in Nepal. The Maoists also talked about it.

The evidence shows that there was free movement of the people of India and Nepal across the boundary between the two countries before the signing of the Sagauli Treaty.[8] After the conclusion of this treaty and the delineation of the Indo-Nepal border, it was decided to continue with the open border between the two countries. An open border was preferred by the British for three reasons. First, to encourage free movement of people across the national boundary as there was the need of Nepali labour to work in tea estates, various development projects, recruitment in the armed forces, etc. Second, there were business and commercial interests. British and Indian businessmen were interested in access to the Nepali markets. Similarly the open border could facilitate the supply of Nepali raw materials, particularly forest-based ones, to India and beyond. Third, the British were also interested in the free access to the Nepal Himalayas by the surveyors, expeditions and treckers etc. After independence, India and Nepal continued with the open border policy in view of the deep-rooted socio-cultural, economic and political relations between the two countries. The Treaty of Peace and Friendship concluded by the two countries in 1950 reinforced the need for an open India–Nepal border in view of various socio-economic provisions that the treaty laid down for the people of the two countries.

Earlier on a system of levying custom duties existed at certain fixed points on border. At present there are 22 agreed custom routes and six immigration points (see Appendix). India has also provided transit to Bangladesh to trade with Nepal through the Phulabari corridor. Occasional conflicts over the border arise between the two countries when the boundary pillars are washed away by floods or removed by miscreants, or are displaced for some other reason. It becomes a matter of concern when no steps are taken for a long time to re-erect them. Once they are placed in their original place, the matter is resolved.

Geographical features

The India–Nepal border is an example of how geography can help in interlinking two countries. The paddy fields, sugarcane fields, orchards, industries, settlements, roads and markets stretch from one side to the other side of the border, thereby making it difficult to recognize the border unless one follows the boundary demarcation pillars. The southern border stretching approximately 1,700 km, is within the Tarai region. It is a humid green belt rich in fertile land. On the Nepali side the area is known as Madhesh and the people of this belt are called Madhesis. The region being above sea level, it has a hot and humid climate in winter and hot in summer. The Nepali area along with the border is within the districts of Jhapa, Morang, Sunsari, Sapteri, Sirha, Dhanusha, Mahottari, Sarlahi, Rauthat, Bara, Parsa, Chitwan, Navalparasi, Rupandehi, Kapilbastu, Banke, Baradia, Kailali and Kanchanpur. The adjoining Indian region consists of Uttarakhand, Uttar Pradesh, Bihar and West Bengal states. The border region is characterized by:

- unnatural frontiers;
- densely populated regions;
- the richest agricultural belt of the Indo-Gangetic plains;
- an urban industrial belt;
- development of transportation and communication facilities;
- easy accessibility to other regions.

In the west of Nepal, the Mahakali river forms a natural border between the two countries. The border region is hilly, covered by dense forests, valleys with scattered population, and small stepped agricultural fields. The western border joins the Uttarakhand state of India and the Darchula, Baitadi, Dadeldhura and Kanchanpur districts of Nepal. Whereas the Kanchanpur district is in Tarai, the rest of the area is hilly and mountainous. In the east, the Mechi river forms a border between India and Nepal. The bordering Indian region belongs to Sikkim and the West Bengal states while the adjoining Nepali area is Taplejung, Panchthar, Ilam and the Jhapa districts of the Mechi zone. This part of the border also has similar characteristics to the western border region.[10]

Thus the geographical features of the Indo-Nepal border are uneven and bear the characteristics of the plains and the hilly and mountainous regions. The border in the plains area is particularly significant from the point of view of multidimensional relations between the two countries.

Border relations: dimensions and issues

The open border has shown various dimensions of close relations between the people of the two countries and has facilitated better understanding and interdependence. These interlinks have become so strong over the years that the open border has become a way of life for the people of the border regions of both sides. It is only the people living in the area who can understand the importance of an open border for them. Some of the important dimensions of the interlinks are referred to below.

Border economic interlinks

The open border provides a large number of economic opportunities and facilities to the people on either side. In fact, for the people of the border region an open border is an asset for many reasons. It is easy to cross the border in search of employment on the other side, and this is more so for the people of Nepal. Because of urbanization and industrialization and ongoing development works on the Indian side, both skilled and unskilled labour is needed and a large number of people from the Nepali side cross the border every day to work on the Indian side. The transport network on the Indian side is quite developed in the bordering Indian towns which provide job opportunities to the Nepali people. During the harvesting season agricultural labour is required on a large scale which is attractive to the Nepalis. Indian agricultural labourers also go to Nepal during the cultivation and harvesting seasons and there they are known as *Dakshinaha* (southerners).[11] The bordering Indian market provides opportunities to the Nepalese for selling their products and for purchasing goods of daily necessity and luxury items, including petrol, kerosene, edible items, cloth, medicines and various other goods. In many areas people of the neighbouring Nepali region make their marriage and festive purchases from the Indian market. In this regard they take advantage of quality and lower cost. The Indian market facilitates the bordering people of Nepal in three ways: availability of goods in case of crisis on the Nepali side, benefit of comparative prices and better market facilities for products including agricultural produce. The open border also facilitates the local entrepreneurs and traders of both sides and they take advantage of comparative prices, consumers and scarcity of goods.[12] In fact, it can be said that there exists a parallel economy in the border region whether it is market, agricultural *Mandis*, etc., and it works despite various rules and regulations on both sides. It is true that there are vested interests and profit seekers: the fact remains that all these economic linkages have survived and flourished because it serves the interests of the local people. The weekly markets (*hat bazaars*) provide an opportunity to the people of both sides to sell and purchase agricultural and livestock products, local handicrafts, etc. Thus, they meet the needs of the poor people of the border areas.

Open border and development of Tarai

The Indo-Nepal open border has played a significant role in the economic development of the Tarai region of Nepal. Developmental activities in Tarai began during the nineteenth century. It had forests, swamps and hot climatic conditions, which made the people of the hill region wary of going there.[13] Therefore, the Nepali rulers encouraged Indians to migrate there and start cultivation. It was the Indian settlers who converted Tarai into a rich agricultural belt. Thus they became the owners of land as well as tenant farmers. This opened up a process of urbanization and infrastructural development in Tarai. Indian traders also began to migrate to the Nepal Tarai region and started business and trade. The industrial development in Tarai began towards the mid-thirties of the last century. Agro-based small and medium-scale industries were also established. In this sector Indians played a pivotal role. The large-scale Indian migration to the Tarai region also promoted its economic interlinks with neighbouring India.

It was towards the mid-sixties that Nepal began to encourage hill migration to Tarai. A large number of incentives were given to the hill settlers in Tarai who wanted to set up business, industries, hotels, etc.[14] Land reforms were introduced and the hill migrants were favoured through policies of land redistribution. However, the hill migration to the Tarai region of Nepal has complicated the socio-economic and cultural fabric of Tarai.

Infrastructural facilities

The open border has provided access to specialized and infrastructural facilities on the Indian side for the Nepalese. The bordering Indian region is much more developed when compared to the Nepali region, particularly in the hilly areas. Choices and alternatives are available to them. The transport network is also well developed on the Indian side. At several places the bordering Nepalese travel via India to visit other parts of their country. A few years back people from Mahendranagar used to travel from Tanakpur to Pithoragarh (Jhulaghat) in India in order to visit Baitadi and places further on in the hill areas. Even though the transport system exists in the Nepali region itself, the roads are not black topped and journeys are tedious. For areas like Darchula district, India alone provides access by road.

For health services, education, entertainment and other such facilities the Nepalese people living in border areas like Darchula, Baitadi, Dadeldhura and Kanchanpur are extensively dependent on India, especially for healthcare and education. Even if these facilities were available on the Nepalese side, they are ill equipped, lack trained staff or are underdeveloped. However, in Tarai, in cases where facilities available in Nepal are cheaper, bordering Indians also avail themselves of these.[15] Special facilities like eye treatments, laparoscopy, general health camps and health programmes organized by charitable trusts on either side are utilized by people of both sides.

Socio-cultural links

The open border, personal contacts, frequent movement of population and their settlement in the Tarai of Nepal particularly, has led to the development of common religion, culture, languages, traditions, social norms, behaviour and pattern of life on both sides. For long there has been a continuous process of Nepalese migrating to the plains of India and the open border has provided an incentive to the frequent flow of people across it. This process of migration of people has accorded such a socio-cultural complex to the entire border region that it is difficult to make a distinction between the people of the two countries. This has encouraged the authorities of both countries to keep the border open. It has also given rise to a sense of cultural overlap across the border, so much so that the concept of a national boundary in the Tarai region has merely a politico legal connotation. The people on either side of the border intermarry. These deep-rooted socio-cultural links and people-to-people relationships have necessitated keeping the border open. In fact, the open border between the two countries has become a symbol of social and cultural continuity.

Political interaction

The people of the two regions do take interest in the political affairs of either side. The general elections are such occasions. They even call upon each other for the support of a candidate or a political party. As a result of these practices borders are closed on the occasion of elections.

Any kind of disturbance in the political relations between the two countries has direct or indirect implications for the people of the bordering region. Since the border is a sensitive issue, it affects the open border most. One significant example of such a situation was the political tangle between the two countries during 1988–89. Relations had become tense over the issue of revision of the treaties of trade and transit. The border was sealed because of the expiry of the treaty and as a consequence the people of the border region had to face economic hardship. It shows how the nature of political relations between the two countries impacts upon the open border.

It is clear from the above discussion that an open border facilitates the people of the two countries in a number of ways. In fact, it can be said that the open border has become an instrument of mutual interdependence more so for the people of the adjoining Nepali region that is less developed and where the people need alternative means of livelihood. These are positive dimensions of the open border. However, there are negative dimensions as well.

Problems

Various problems and issues relating to the open India–Nepal border are now discussed.

Social hazards

Because of closer social interaction between the people of the two regions facilitated by the open border, socio-political tensions of the one side have direct impact on the other side of the border. The social conflicts, communal tensions and local disputes have their repercussions on both sides. Since their caste, religious and linguistic customs are similar, any social upheaval on the one side aggravates the situation on the other.

Anti-social activities and criminal acts, etc., are a big problem in the border region. In fact, the open border is considered to be a safe place by criminals. After doing a criminal act on one side they can easily take shelter on the other side of the border. Various criminal and lawless activities, like theft, robbery, kidnapping, murder, looting, threatening of businessmen and use of anti-social elements in land grabbing and confiscation of private properties are included. This often results in incursion of the border by police and security forces in search of criminals which can cause political tensions between the two countries. The trafficking of women across the Indo-Nepal border is a serious problem. The open border is misused to send Nepali girls to India for work as sex workers in brothels. A large number of people on both sides of the border are said to be involved in this racket. It has resulted in the spread of sexually transmitted deceases like HIV and AIDS.

Unlawful activities

The smuggling of arms and explosives through the Indo-Nepal open border has become a serious problem. It is true that there are check posts at different points all over the border, with the police, customs officials, revenue officials, immigration officials and security forces being deployed on both sides to control unlawful activities. Yet, one would agree that the management of the open border is very complex and cumbersome.[16] Both the imported and country-made arms and explosives are smuggled across the border. Another problem is that of drug trafficking. Nepal is considered safe for the illicit drug trade. Apart from trading of *Ganja* and hashish, synthetic drugs like ecstasy pills, amphetamine, methamphetamine, etc., are also smuggled.[17]

The open border has equally given rise to security problems. In recent years terrorist activities have increased. It is difficult to control these activities due to the porous nature of the border. Nepal has been considered a safe haven by the terrorists and intelligence agencies such as ISI (Inter Services Intelligence of Pakistan). It is easy for them to cross the border and India's worries have indeed grown with this increased use. The large amount of aid coming to the Tarai region of Nepal from Islamic countries has promoted the extremist activities which are further aggravating the situation and the security problems for India.

Maoist insurgency and problem of security

The Maoist insurgency in Nepal has added new dimensions to the security risks as the insurgents began taking advantage of the open border.[18] The security implications of the Maoist problem caused India to express its concern over the situation in Nepal. The problem has security threats for India in several ways.

- India's perception has been that it cannot remain unaffected by domestic turbulence in Nepal largely due to the open border between the two countries.
- The Maoists use it to take shelter in Indian territory at their will. They even enter it for medical treatment of their combatants. There have been frequent visits of their leaders into the Indian territory for consultations, meetings, etc.
- There large-scale Nepali migration to the bordering Indian districts in Uttarakhand and Uttar Pradesh in the wake of insurgency in Nepal.
- The links between the Nepali Maoists and the revolutionary groups in India such as the Naxalites, People's War Group, etc., are well known. Needless to say that the open border made this interaction easier and frequent.
- The Maoist insurgency in Nepal also provides an edge to a number of

illegal activities including supply of illegal arms and terrorist activities in India and along the Indo-Nepal border.

It may also be added here that in the bordering areas of Uttarakhand and Uttar Pradesh terrorist activities increased during the period of the Maoist insurgency in Nepal. Thus the open border becomes a security risk for both countries due to its porosity.

Border demarcation

The problem of border demarcation has existed since the Treaty of Segauli. In 1829 it was agreed that border pillars would be re-erected whenever they were destroyed. The border pillars were destroyed with an intention to grab land by the people of both sides or by accident. This created confusion over the line of demarcation and gave rise to disputes between the two countries. Secondly, rivers like Kosi and Gandak sometimes change their course and destroy the existing border pillars, which also led to problems in border demarcation.

Border disputes have serious implications for relations between the two countries. In the 1960s the Susta border dispute came to prominence. The opposition political parties in Nepal raised the issue of displacement of border pillars in the Tanakpur dam area along with the Tanakpur barrage controversy. Nepal had also raised the Kalapani issue by claiming that the Kalapani area belonged to it. This region forms a narrow neck along the Mahakali river in Pithouragarh and Dharchula districts of India and Nepal respectively. It is a tri-junction of India, Nepal and Tibet and Indian paramilitary forces have been guarding this region since the Chinese attack in 1962. The issue remains to be settled and Nepal has often charged India with the unilateral replacement of border pillars.[19]

Problem of smuggling

Smuggling across the Indo-Nepal border is well known. Only its nature and contents have changed over the years. Today this border is considered a smuggler's paradise. There is no definite data about illegal trade taking place through the border but it has been estimated that the value of goods smuggled between India and Nepal may be as high as eight to ten times the official bilateral trade.[20] The complete freedom of currency movement between the two countries facilitates illegal border trade.[21] Generally there have existed six patterns of smuggling across the border.

- Indian manufactured goods exported to Nepal under the 'quota system' are cheaper in comparison to their prices in India. Hence, many a time as soon as these goods reach the Nepali market, they are smuggled back into India in order to earn more profits.

- Some donor countries provide aid to Nepal in the form of consumer goods instead of cash grants. Since there is no or little internal market for such goods, these goods, after reaching the Nepalese market, are smuggled into India.
- The contradictory export-import policies of the two countries also provide scope for smuggling across the border. For instance the Nepalese Government provides special concessions to its traders for exporting certain items to Third World countries, in order to give a boost to its exports. Sometimes the Government of India raises export duties on similar products so as to ensure the proper supply of these goods to its industries and discourage exports; these goods are smuggled from India to overseas countries via Nepal.
- In the 1960s encouragement for smuggling came through two schemes followed by the Nepalese Government. 'Exchange Entitlement' from export earnings for import of luxury goods was introduced in 1964 and the 'Gift Parcel Scheme' was introduced in 1967. Traders with vested interests imported Third World goods with the intention to smuggle them into India, thereby exploiting these schemes. It is believed that the high tariff structure in India and lower tariffs in Nepal, and lack of common policies in the case of Third World goods, encouraged smuggling across the border.[22]
- When the prices of certain goods, particularly petroleum products differ, they are smuggled across the border.[23]
- Smuggling of surplus agriculture produce from the Nepal Tarai into India is yet another issue. Actually, easy transportation of goods into the Indian plains, better marketing facilities and higher prices on the Indian side encourage the farmers and landlords of the Nepal Tarai to smuggle their agriculture produce to India.

An illegal trade of Rs. 30 billion takes place between the two countries annually.[24] With the rise in gold prices, smuggling of gold has increased along the border;[25] 2.5 kg of gold is reported to be smuggled every day.[26] Despite several efforts made by the governments of the two countries to check smuggling across the border, people on both sides continue with these activities. The Agreement on Cooperation for Checking Unauthorized Trade is not being implemented properly.

Uncontrolled migration

The open border has greatly facilitated migration between the two countries and this has been viewed as something similar to unrestricted entry rules generally applied to transboundary ethnic groups.[27] There is no quota system to determine the number of migrants crossing the border.[28] In fact, the number of people who cross it daily is so large that it is not possible for the authorities to regulate them on either side. These migrants have wide-ranging interaction with the people on the other side. The consequences of such migration have been felt more in the Tarai region of Nepal than elsewhere[29] as it has considerable bearings upon the population growth in the Tarai as well as the socio-religious character of the region.[30] Nepal, having less absorption capacity and fewer resources, seems to be more concerned about the problems of international migration and is looking for certain measures to discourage it. However, these measures have not been effective.[31]

Dependency syndrome

The open border increases Nepal's economic dependence upon India to a considerable extent.[32] The people of the Tarai region are more dependent on the bordering Indian market and traders for their daily necessities and employment so that the economy of the Tarai is viewed as an adjunct to that of India.[33] It may also be pointed out that the overall structure of the Indian economy has tremendous influence over the Nepalese economy. India's economic policy measures influence the Nepalese economy as well, and Nepal is also prompted to take similar measures with the changes in the Indian market economy. This type of interaction has become multiplied due to the open border.

Tarai agitation and the open border

The open border has directly or indirectly added to the problems faced by the people of the Tarai region of Nepal. This region actually forms an open border with India, not only due to its geographical similarity with the bordering Indian region but also as a result of deep-rooted socio-cultural and economic interlinks between the two regions. The economy of Tarai has been viewed as an extension of the Indian economy, largely due to the nature of economic interdependence among the people of the bordering region. The people of Tarai have more socio-cultural interaction with the neighbouring Indians compared to the people of the Nepalese hilly areas. In the past, politically motivated violence and activities against the Kathmandu-based ruling elite had created fear and suspicion in their minds about the people of the Tarai region. They were considered more loyal to India than to Nepal.[34] There have also been problems in the emotional integration of the people of the Tarai with those of the hills,[35] as the people of Tarai nurse a feeling of relative deprivation. There is also the problem of stateless persons in Tarai.[36] The issue of autonomy of Tarai has become very heated during the past few months. In fact, the people of Tarai have been demanding adequate representation in the political and administrative structures of the country. There has indeed been serious discontent among them against the discriminatory attitude of the hill-dominated ruling elite. The Maoists took advantage of the widespread discontent in Tarai and assured them autonomy in the new political organization of Nepal. The Tarai people are desperately looking for the realization of their demands and their frustration has burst into a form of serious agitation in Tarai since January 2007. The demand for autonomy and federal structure of the Nepalese state has become an important issue. The groups that are leading these agitations in Tarai are splinter groups of the Maoists. Disturbances in the Tarai also have political, economic and social implications for India. The leaders of Tarai also look towards India for support. They take advantage of the open border to run into the Indian side whenever they feel pressure from the Nepali authorities. The border region being sensitive, peace and stability on one side of the border is equally important for the other.

Implications for bilateral relations

Problems associated with the open border as discussed above have serious implications for bilateral relations between the two countries. The controversy over displacement of border pillars has often been raised. The problems of migration and of smuggling have been a bone of contention. The Nepalese rulers have often raised the issue of Indian economic hegemony in the Tarai. Criminal activities across the border ultimately lead to tensions in bilateral relations. The terrorist trespassing is a matter of concern for both sides. No less serious is the problem of Nepali citizenship to the people of Indian origin in Tarai.

The implications of these problems in bilateral relations of the two countries depend on two factors: Whenever the political understanding has broken down between the two countries, these issues have been raised. At times, the issues related to the open border are raised in order to score political gains against each other. It should also be made clear that these issues have the potential to generate political confusion and create deadlock in the relations between the two countries.

Differing perceptions

There are differing perceptions over the issue of an open border. It is true that this has existed for a long time because of the common approach of the two countries. However, in recent years, the Nepali ruling elite has started viewing the open border as a cause of problems and tensions. It is often viewed as a source of dominance from the south, whether it involves migrations, economic activities or political issues. The open border issue has at times been used as a political tool.

During the period of Panchayat rule in Nepal, the ruling elite of the country strongly believed that the open border was a source of many tensions. The government was particularly concerned over the Indian immigration to Nepal. It gave serious thought to the regulation and even the closing of the border. The idea of controlling it still persists in the political and administrative circles of the Kingdom and also a section of the Nepali intelligentsia. However, the perception of the Tarai people is quite different. They allege that the perception of the Kathmandu based ruling elite and intellectuals is removed from reality. They believe that the open border not only needs to be maintained but also further strengthened. This view is supported by the Nepali academia belonging to the Tarai region.

From the Indian point of view, two aspects are important:

- The open border with Nepal is in accordance with the spirit of the 1950 Treaty of Peace and Friendship. Both are supportive to each other. So long as one provision exists, the other has its relevance.
- India's concern would be over the increasing security threats along the border. It is the misuse of the open border which needs to be checked with strict vigilance. It would not affect India much if the border is closed. India seems to have faith in the existing provisions. Moreover, an open border is not as debatable an issue in India as in Nepal. But the elitist and popular perceptions relating to the open border are very different in Nepal.

What should be the remedy to these problems then? An easy answer would be to close the border. Many in Nepal have advocated this viewpoint. But it is too simplistic a solution to such a complex phenomenon. A closed border would bring more hardships to the people of the border area. In the context of increased terrorist trespassing, large-scale migration and smuggling, it is necessary to take certain long-term measures. The realities of an open border are to be reckoned with. It is only by resolving these constraints that the open border can play a role in the development of beneficial bilateral relations. It may be pointed out here that the provision of free access to each other by the people of the two countries, economic inter-dependence, etc., go along with the spirit of the 1950 Treaty of Peace and Friendship concluded between them. While the relevance of the 1950 treaty cannot be questioned, it is necessary to look into the problems that are associated with the open border. Effective measures need to be taken in order to resolve these problems. However, in taking up any measure the sensitivities of the people of the two countries living across it need to be borne in mind, and also the fact that a controlled border is not much different from a closed border. Any measure in this regard has to be a joint programme, as unilateral action would not help in resolving the problem. An effective border management without harming the popular sensitivities and patterns of interaction and measures to protect the open border against unlawful activities is needed. In fact, many of the problems have occurred because the border is not properly guarded against such activities. In recent years, due to increasing security risks, both countries have taken several steps to minimize the problems encountered along it:

- India and Nepal have agreed on the extradition of criminals and wanted persons;
- verification of personnel and exchange of information about criminal activities has been agreed;[37]
- it has been agreed to fence the border up to 2 km at each custom check point and there will be approximately 9 metres of no man's land on either side along the fenced border;
- both countries have agreed to share information on illegal trade;
- a joint standing committee has been constituted to look into problems arising due to floods in the border area;[38]
- The two countries have agreed to curb illegal trade.[39]

Many other measures can be taken to resolve the problems. What is more important is that these measures have to be implemented and made effective. It is also necessary that individual countries should restrain themselves from taking any unilateral decision on the border issue, as it may hurt the sensitivities of the other country.

Future prospects

There are different perspectives regarding an open border. Some scholars believe that it is a unique provision and should be maintained. They also argue that other countries of the region should follow the trend, keeping in view the deep-rooted socio-cultural links between the countries. This view is shared by the people, particularly those of the bordering region. It is also argued that the process of globalization encourages the opening of international borders. But it is known to be a source of many problems. The way it has facilitated terrorist activities, illicit trade, trafficking of drugs, women and arms and posing serious security challenges, is an acknowledged fact. There are people in Nepal who support the idea of closing the border, but history bears testimony to the fact that the one between India and Nepal has survived despite many odds. The issue is not to ignore the problems and challenges but to resolve them with mutual understanding and cooperation. The main issue is, therefore, of proper border management.

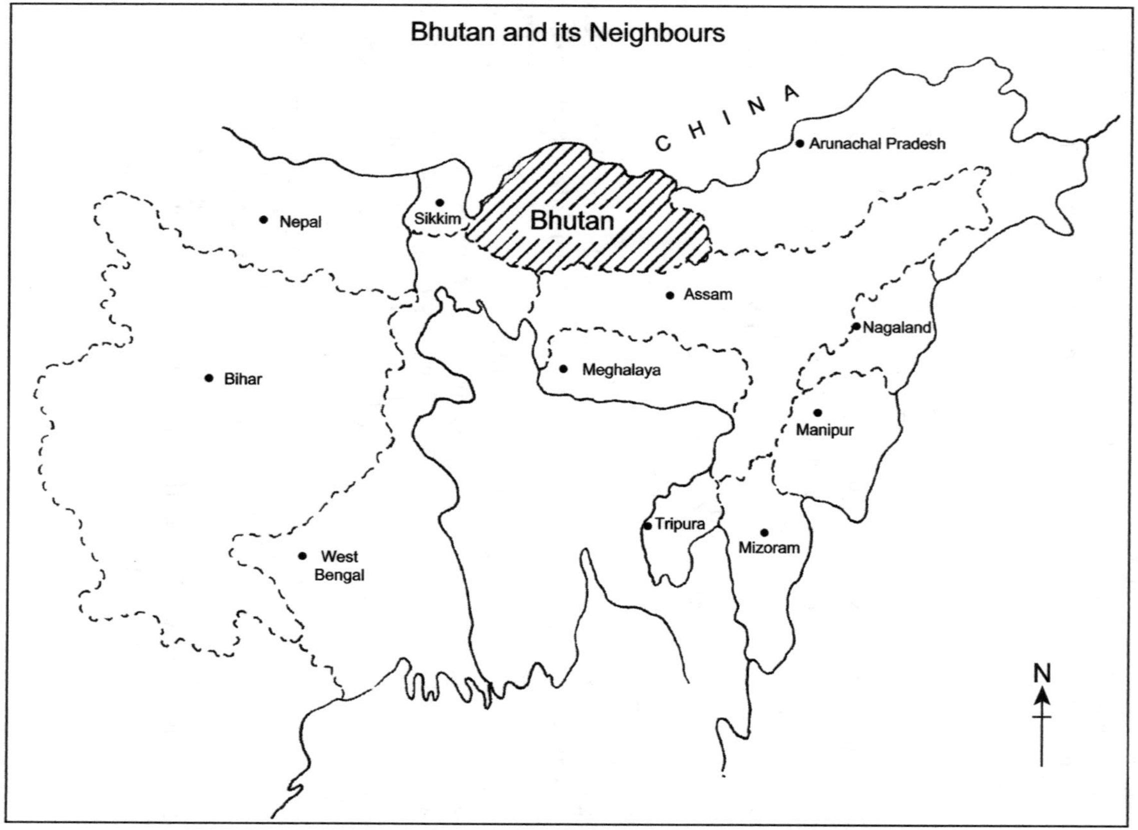

Bhutan and its Neighbours
CHINA
Arunachal Pradesh
Nepal
Sikkim
Bhutan
Assam
Nagaland
Bihar
Meghalaya
Manipur
Tripura
Mizoram
West
Bengal
N

Figure 11.1

11 Indo-Bhutan relations

Strategic perspectives

Rajesh Kharat

Geostrategic location

Geographically speaking, Bhutan is a very small country, covering an area of approximately 18,000 sq. miles or 47,000 km^2. It is relatively compact with a maximum north–south distance of 170 km and a maximum east–west distance of 300 km. The geographical location and land-locked nature of Bhutan isolates it from the rest of the world. On the north, it is bounded by Tibet, whereas on the south it is surrounded by the plains of the Jalpaiguri district of West Bengal, and the Golpara, Kamrup and Darrang districts of Assam. Bhutan is sandwiched between the Chumbi Valley of Tibet, Sikkim and Darjeeling in the west and the Kameng district of Arunachal Pradesh in India, on the east side.

Since Bhutan is land-locked and sandwiched between two Asian giants – India and China – it maintains its relations and contacts with the rest of the world through Indian territory. For Bhutan, Calcutta is the nearest airport as well as its nearest seaport. By road, it is only 750 km away from Phuntsholing, its main gateway town.

Inaccessibility is one of the characteristic features of the country. Geographical factors such as thick forests, high mountain ranges, adverse climatic conditions such as severe winters and heavy rainfall, all make Bhutan inaccessible from the east, west and the north. The mountainous terrain also does not permit the country to have railway lines. The only possible transport is by air. Another feature of this country is its sparse population, not exceeding 10,000 to 15,000 persons in any place. The mountainous region and thick forests contribute to the isolation both within and outside Bhutan.

Yet Bhutan has maintained a steady contact with its immediate neighbours, having had close contacts with Ladakh, Lahul, Spiti and Kinnaur in the west, Tibet in the north, and Sikkim, Cooch Behar and northern Assam. Bhutan not only maintains trade relations with this region but has formed an ethnic and cultural unit over the years. Its geo-strategic location has also influenced Bhutan's political and economic relations with other countries. The events in Tibet of 1958–59, and on the Indo-Tibetan border in 1959, resulted in widespread resentment among the Bhutanese, as it caused a considerable setback to its economy due to disruption of Bhutan's trade with Tibet. The only option remaining with Bhutan has been to depend on India for its trade relations with the rest of the world. After the Sino-India war of 1962, the geo-strategic location of Bhutan played a significant role in its external relations, as it forms the external rim of the Himalayan mountain region of India and the Indian states of bordering West Bengal, Sikkim, Assam and Arunachal Pradesh.

Although agriculture and animal husbandry are the main features of Bhutan's economy, its production of food grains is insufficient due to the natural environment and non-availability of infrastructure like irrigation, seeds and new scientific methods. The mountainous terrain becomes a hurdle for smooth internal transportation and communication. Ninety five per cent of the population is dependent upon animal husbandry and agriculture. Bhutan's forest wealth is yet to be tapped to the fullest extent. There are minerals such as limestone, dolomite, coal, gypsum, graphite, copper, lead, zinc, marble, slate and talc. India is helping Bhutan to exploit these resources. Bhutan has only two sources of fuel and power, namely coal and hydro-electricity. Since coal reserves are limited in quantity as well as quality, Bhutan imports almost all its coal from India. The other source of energy, hydroelectric power, depends on its rivers, the Torse, the Raidak and the Manas. However, only a small portion of this enormous hydropower potential has been tapped, because of inaccessibility, and lack of capital investment and technical know-how for development. Bhutan and India signed a pact in September 1961 to harness the Jaldhaka river and to purchase the power thus generated from Bhutan for India's state of West Bengal.

The developmental process in Bhutan also suffers from the lack of the means of communication and efficient transport. Moreover, the industrial base is very limited, being confined mainly to the cottage level. As such its economic relations, including its external trade, are 90 per cent linked with its immediate neighbour India. Bhutan's telecommunication links with the outside world run through India. Thus, being a small land-locked country, the economic activity of Bhutan is dictated by its geographical situation.

India's Bhutan policy (1949–2006)

Bhutan's first contact with British India was made when its raids against Cooch Behar forced the latter to ask the East India Company for help in 1772.[1] This contact gradually evolved into a closer bilateral relationship especially after 1907, when Ugyen Wangchuk became the King of Bhutan. To contain British India's imperialist policy in the region, the Chinese announced their claim over Bhutan and started interfering in its internal matters. British India, therefore, decided to revise the 'Treaty of 1865'[2] to protect Bhutan from the Chinese influence.

The Political Officer in Sikkim, Charles Bell, recommended to the Government of British India that to counter the Chinese advances, Bhutan should submit all its disputes with foreign powers for consideration to the Government of British India. In this manner, British India influenced Bhutan's external relations.[3] Finally, a new treaty was concluded on 8 January 1910 between Bhutan and British India at Punakha, whereby the Government of Bhutan agreed to conduct its foreign relations under the guidance and advice of British India.[4]

From the Bhutanese point of view, the 1910 Treaty of Punakha not only changed the political history of Bhutan but also the social and economic life in the country. Maharaja Ugyen Wangchuk of Bhutan signed this treaty for three reasons. First because he wanted to protect Bhutan from China's expansionist policies; second, it was not possible for Bhutan to maintain a separate political entity without the help of British India; and third, he felt this was the way to modernize Bhutan and bring radical changes into the economic system of the country with British India's assistance. Ugyen Wangchuk thus not only ensured the autonomy of Bhutan, but also kept it free from foreign influence.[5]

Soon after independence, India's then Prime Minister Jawaharlal Nehru, assured all neighbouring countries that his government was bound by the obligations of the Treaties and agreements of former British India. Bhutan thus sent its delegation headed by Raja Sonam T. Dorji on 23 April 1948[6] to discuss their relations with the new independent India.[7] When by the summer of 1949 the Communists controlled almost all parts of China's mainland and established their government,[8] it was seen as a threat for both Bhutan and India. Thus security considerations compelled the two countries to maintain close and friendly relations with each other. India and Bhutan held negotiations, which lasted until August 1949. Whereas India gave an assurance that it would respect Bhutan's independence, Bhutan promised to maintain the same relationship which existed with the British.[9] India also agreed to return the area of the Dewangiri hill strip, an area of 32 sq. miles (83 km^2) in eastern Bhutan (Bhutan had renamed it as Deothang) after Bhutan agreed to revise the essential provisions of the treaty of 1910.[10] Bhutan and India signed a new treaty at Darjeeling on 8 August 1949.

The Indo-Bhutan Treaty of 1949

From the Indian side, this treaty was ratified at the highest level by Jigme Wangchuk, the Maharaja of Bhutan, on 15 September 1949 at Tongsa and President C. Rajagopalachari on 22 September 1949 at New Delhi.[11] The treaty conceded the independent and sovereign status of Bhutan and played a vital role in promoting and fostering friendly relations with India.[12] Article 2 of this treaty says that:

> The Government of India undertakes to exercise no interference in the internal administration of Bhutan. On its part, the Government of Bhutan agrees to be guided by the advice of the Government of India in its external relations.

Thus, India assured Bhutan of its non-interference in its internal affairs. From the Bhutanese point of view, the incorporation of Article 2 in this treaty has a different meaning altogether. So far as the first part – 'non-interference in the internal administration of Bhutan' – is concerned, it clarified the sovereign and independent status of Bhutan. However, Article 2 of the treaty put restrictions on Bhutan from extending its relations beyond India. Bhutan realized the importance of this treaty on account of the rise of Communist power in China and its claims on Tibet, Bhutan and Nepal,[13] which was brought home forcefully after the forced incorporation of Tibet in 1959. Hence, the treaty of 1949 played a significant role in the formulation of the foreign policy of Bhutan.

From the Indian point of view, Article 2 of this treaty has its own significance. In the first place, the treaty was a logical continuation of the British–Indian arrangement. Bhutan is strategically located on India's northern border. Nehru had emphasized that India should take care not to get involved in the internal squabbles of Bhutan (which he thought, in turn, might encourage China or other foreign influences within its territory).[14] In a private meeting with B. M. Kaul, Nehru stated.

> how important it was from India's point of view to strengthen Bhutan's friendship in view of her key position on our border and how we must do everything possible to help her. He also said we must treat smaller countries like Bhutan as our equals and never give them an impression that they were being 'civilized' by us.[15]

Thus, from the defence point of view, India benefited from this security arrangement. As an ally of India, Bhutan provides a natural barrier to protect the Himalayan frontier of India by preventing the Chinese from entering into India. Bhutan's loyalty to India is dictated by the fact that the land-locked position of Bhutan makes it dependent on India for trade, transit and contact with the outside world. Moreover, there are mutual security arrangements like the Indian Military Training Team (IMTRAT) in Bhutan.

According to Article 3 of the treaty, the Government of India increased the compensation grant as a consolidated amount of Rs. 5 lakh to be paid annually to Bhutan. Article IV established the principles of equality in the bilateral relationship between Bhutan and India. In support of the principle of equality, the Government of India agreed to return an area of thirty-two set miles called Dewangiri to Bhutan. Article 5 of this treaty established free trade and commerce between India and Bhutan, with India providing free access for trade to Bhutan through its territory. By this treaty, Bhutan as a sovereign state entered into a special relationship with India, which after a few years developed into friendly assistance for the former's economic development. Thus, Indo-Bhutanese friendship became the cornerstone of Bhutan's foreign policy. The Indo-Bhutan Treaty of 1949 strengthened the position of Bhutan at home as well as in the outside world. Meanwhile Jigme

Dorji Wangchuk succeeded as the King of Bhutan in the hereditary system of monarchy.

In 1958, Indian Prime Minister Jawaharlal Nehru visited Bhutan. During his visit, Nehru reiterated his government's cooperation with Bhutan and expressed his hope that Bhutan would not misunderstand Indian policies. Nehru said:

> Some may think that since India is a great and powerful country and Bhutan a small one, the former might wish to exercise pressure on Bhutan. It is, therefore, essential that I may make it clear to you that our only wish is that you should remain an independent country choosing your way of life and taking the path of progress according to your will. At the same time, we two should live with mutual goodwill. We as members of the same Himalayan family should live as friendly neighbours helping each other. Freedom of both Bhutan and India, should be safeguarded so that none from outside can do harm to it.[16]

This visit provided an excellent opportunity to have meaningful bilateral dialogue between India and Bhutan concerning various issues, including the economic and social development of Bhutan. Nehru asked Bhutan to become modernized with India's help and initiated the idea of Bhutan's five-year development plans. Nehru also discussed with the Maharaja the possibility of establishing a road link between the two countries.

Whereas India wanted to maintain the independence of Nepal and Bhutan in order to restrict Chinese influence in that region, India was more concerned with Bhutan stating, 'because of the Buddhist cultural background it shared with Tibet, Bhutan may have seemed to be even more vulnerable to subversion than Nepal'.[17] To quote Leo E. Rose, 'by and large, the Bhutanese shared the Indian Prime Minister's concerns over the strategic implications of Chinese controlled Tibet'.[18]

After his visit to Bhutan in 1958 and realizing the importance of Chinese designs in the Himalayan region including Bhutan, Prime Minister Nehru declared in the Lok Sabha on 28 August 1959 that 'the protection of the borders and territorial integrity of Bhutan was the responsibility of India' and that 'India would consider any aggression on Bhutan as aggression on India'.[19]

Chinese takeover of Tibet

The Chinese suppression of the Tibetan revolt in 1959 caused apprehension in Bhutan. During this time, many Tibetan refugees entered the Bhutanese territory without permission. Also, the flight of the Dalai Lama to India and the assemblage of Chinese troops near Bhutan's northern border alarmed the Bhutanese.[20] The developments in Tibet compelled Bhutan to be more cautious and Prime Minister Jigme Palden Dorji visited India in August 1959 and 'sought a written guarantee of Indian support in the event of a Chinese attack on Bhutan'.[21] On 15 September 1959, the King of Bhutan visited India, assuring it that his government had agreed to be guided by India in its foreign policy. The king further said, 'We found India a friendly neighbour and we have no reservations in strengthening these ties.'[22] Thus, Bhutan developed a closer strategic relationship with India.

However, in May 1960 there was some misunderstanding in Bhutan about a map, which was published by the Government of India. Bhutan drew India's attention to the fact that the map had not shown the Indo-Bhutanese boundary as an international boundary.[23] Thin Ley, the Speaker of the Tshogdu (Bhutan's National Assembly), conveyed to the Indian Government that the members had expressed serious concern about this common boundary line and demanded that it should have been shown thicker as an international boundary line. The members, however, were convinced that India was responsible for Bhutan's security.[24] This misunderstanding was removed later with the signing of the Indo-Bhutan Border Agreement in 1973, when the Government of India agreed 'to show Bhutan in a different colour on its maps'.[25] Thus, Bhutan always remained alive to the necessity of safeguarding its independence and asserting its sovereignty within the framework of the Indo-Bhutan special relationship.

In January–February 1961, the King of Bhutan Jigme Wangchuk visited New Delhi to maintain friendly relations with India, and especially to discuss the matter of Bhutan's economic and defence needs. During the talks, the Bhutanese delegation sought Indian help against any possible Chinese attack. In response, the Indian Government, with a moral obligation to protect Bhutan, introduced IMTRAT in Bhutan with the objective of providing military training to the Bhutanese people and to help Bhutan to build up the Royal Bhutan Army for guarding its borders against any possible Chinese threat. During this visit, for the first time, Bhutan's flag was flown along with the Indian flag. Thus, Bhutan maintained special relations with India.

At the economic level, several aid agreements were concluded between the two Governments whereas politically, India helped Bhutan to project its independent identity in international affairs.

Sino-Indian War of 1962

The Chinese attack on Tibet in 1959 and the Sino-Indian war of 1962 'aggravated Bhutan's sense of insecurity and suspicions about China'.[26] This war certainly made Bhutan more alive to the need for its independent existence and for its survival between the two giants. During the 1962 war, some important areas of the Himalayas, for example Ladakh and all areas near the kameng district of the North East Frontier Area (NEFA) which borders on eastern Bhutan, became the battlefield. This war affected the Bhutanese perspective towards both China and India. 'The Royal Government sought to maintain strict neutrality, at least formally, but it was only too aware that its close alignment with India made Bhutan a vulnerable target if the Chinese were disposed to move against it.'[27] In September 1962, China controlled a few strategic frontier posts and forced Indian troops to retreat to the Indian frontier at Se La, close to Tashingang, Bhutan's border. Around this time, some Indian troops crossed into the Bhutanese border and took refuge in the Dzongs of Twang and Tashingang, in eastern Bhutan.[28]

According to the Indian military personnel, the Bhutanese Government had shown sympathy, providing them with medical care and hospitality. Some troops suffered from frostbite and severe cold and were helped by the local Bhutanese. 'The King had ordered shelter and food for them, risking Chinese displeasure.'[29] However, according to Neville Maxwell, the Bhutanese Government later on lodged a protest to the Indian army regarding the entry of these troops. Stating that Bhutan was a sovereign country, it turned down India's offer to station troops for defence purposes.[30] The Royal Government of Bhutan pointed out that the Indo-Bhutan Treaty of 1949 was meant for guiding its foreign policy and was not a defence treaty.[31] It is quite possible that the Bhutan Government had changed its stand from a soft line to a hard line towards the Indian personnel due to the fear of a possible retaliatory Chinese threat, or perhaps from the need for not annoying the Chinese Government. In this manner, Bhutan sought to keep a safe distance from India.

This war weakened India's position in the subcontinent as well as in the eyes of the Bhutanese people. Many in Bhutan expressed doubts as to whether India would be able to protect them if China repeated its Tibetan exercise. The Bhutanese felt insecure because of the Sino-Indian war. Therefore, the post-1962 war period compelled Bhutan to re-evaluate certain aspects of its foreign policy. As a result, the elite class of Bhutan proposed to adopt the model of Nepal's foreign policy, which was 'based upon equal friendship with India and China, with a balance of power strategy'.[32] But the Royal Government of Bhutan did not share the perception of the elite class and rejected the Nepali foreign policy model as not being feasible for Bhutan due to its strategic and political situation. Bhutan decided not to adopt the Nepali model of equidistance between China and India but to opt for closer links with India. According to Leo Rose, the Bhutanese foreign policy strategy was based upon the assumption that New Delhi would not obstruct the gradual expansion of Bhutan's relations with the outside world if this could be accomplished without undermining India's regional and security interests.[33]

At the same time, the Sino-Indian war of 1962 had shaken Bhutan's confidence in India's capability to defend it against Chinese attack. It therefore demanded a revision of the Indo-Bhutan Treaty.[34] In deference to Bhutan's wishes, India decided to encourage Bhutan to develop links with the outside world. India sponsored Bhutan as an observer at the 14th meeting of the Consultative Committee of the Colombo plan for co-operative economic development in South and South East Asia in Melbourne (Australia) on 19 November 1962.[35] India backed Bhutan for membership of the Universal Postal Union in September 1969, and of the United Nations in 1971.

Moreover, in 1973, India sponsored Bhutan as a member of the Non-Aligned Movement (NAM).

Assassination of J. P. Dorji in 1964

There were two major events which resulted in a setback to the cordial relations between India and Bhutan. First, the assassination of Bhutan's Prime Minister Jigme Palden Dorji in April 1964[36] resulted in a crisis within the government and posed a possible threat of civil war.[37] Secondly, there was an attempt on the life of Druk Gyalpo in July 1965. It was reported in the Nepalese press that Indian officials had masterminded the conspiracy to assassinate Jigme Palden Dorji and there were allegations in Bhutan that the commander of IMTRAT was behind the conspiracy.[38]

However, Nari Rustomji (the then Indian adviser to the Royal Government of Bhutan) was quick to deny these allegations. He stated that there could be other reasons for the assassination, such as Dorji's constant support for Indo-Bhutan friendship and Bhutan's alignment with India.[39] After Jigme Dorji had invited Nari Rustomji to become the Indian Adviser to Bhutan, there were allegations of Indian interference in the internal matters of the country. It created a hostile attitude among the local people;[40] they did not like Indian influence in domestic politics or any introduction of modern Western concepts. Be that as it may, it was a well-planned propaganda against India, managed by those who were involved in this incident and had fled to Nepal. The King of Bhutan was categoric in his rebuttal of such allegations. 'This incident has been of a purely internal nature and there is absolutely no foundation in the allegations that the developments have been engineered by the Government of India or any foreign Government.'[41] Druk Gyalpo went out of his way to reassure the Government of India that Bhutan's policy towards India would not change.

However, after this incident, the role of the Indian adviser was minimized and when Rustomji retired, no successor was appointed as the Tshogdu abolished the post. This also led to the abolition of the post of the Bhutanese agent in New Delhi. However, the Indian Political Officer in Sikkim remained officially accredited to Bhutan.

In January 1965 the then Prime Minister of India, Lal Bahadur Shastri, had a meeting with the King of Bhutan to clear the air of suspicion. He discussed various aspects of the Indo-Bhutan relationship and declared that 'Bhutan is an independent country and we have always accepted that.'[42] As a reciprocal gesture, the King of Bhutan visited India in May 1966 and said:

> I am deeply touched by the sympathy and understanding with which the Government of India views our problems. The help and advice furnished to us by the Government of India are of great value to us and are appreciated by my Government and people.[43]

In February 1968, when the king revisited New Delhi, Dr. Zakir Hussain, the

President of India, told him that the Government of India expected that the 'Bhutanese people will come to regard India as a second home away from home'. In response, the king while referring to Indo-Bhutanese relations said, there was 'no misunderstanding, no dispute, no argument between our two countries'.[44]

It was the time when both countries felt the need to upgrade mutual diplomatic relations. It had also become necessary for Bhutan to extend its international contacts. In 1968, both the governments of India and Bhutan finally decided to appoint a Special Officer of India at Thimpu, to act as a link between the two governments to cover all matters of mutual interest.[45] He was also required to coordinate, expedite and facilitate the implementation of Indian-aided projects in Bhutan. It is worthwhile to note that when India had proposed to post an officer at Thimpu, way back in 1967, the same members of the Tshogudu had criticized the proposal.[46] They had argued that an officer might exert an unwarranted influence in the internal affairs of Bhutan.[47] The official with the title of 'Special Officer to Thimpu' had the restricted role of looking after diplomatic functions only. Initially, Bhutan was reluctant to have a mission in New Delhi, but it changed its position after it became a member of the United Nations in 1971. Thus, the two countries agreed to exchange each other's diplomatic representatives in 1971.[48]

In 1972, Bhutan opened its permanent mission to the United Nations at New York.[49] This provided Bhutan with an alternative channel to communicate with the outside world, while at the same time it reduced the importance of its mission in New Delhi. With these two missions, Bhutan tried to maintain and regularize its relations with foreign countries. Political relations between India and Bhutan remained friendly and stable, particularly in the late 1960s and early 1970s. The King, who was the chief architect of the India–Bhutan friendship, visited India several times.

The Merger of Sikkim with India and its aftermath

Situated at the western border of Bhutan, Sikkim shares its religion and the Mahayana Buddhist tradition with Bhutan. Between 1973 and 1975, several stormy incidents took place in the internal politics of Sikkim. These led to the abolition of the monarchy and the merger of Sikkim with the Indian Union in 1974. Bhutan described this event as a 'loss of cultural identity of a neighbouring Himalayan Kingdom'.[50] The Nepali Sikkimese population had sought Indian support for providing a new political system in Sikkim and have it merged with the Indian Union.[51]

Apart from its apprehensions about the loss of monarchy in Sikkim, Bhutan was more concerned over the rising political consciousness among the Nepali Sikkimese, which posed a threat to the ruling monarchy of Bhutan.[52] Bhutan was also worried that the fate of Sikkim may well befall Bhutan sooner or later.

One can understand Bhutan's concerns regarding the Sikkim episode and its passive fall-out affecting Indo-Bhutan relations. But it is also a fact that it was not the 75 per cent of the Nepalese population of Sikkim that forced the merger, or affected a change in the ruling pattern. It was the discrimination and economic disparity created by a feudal system for self-preservation that actually led to their agitation.[53] Besides, the situation of Bhutan and Sikkim had been qualitatively different both in legal and political terms. Before 1947, Bhutan was not a part of British India, unlike Sikkim. Consequently, independent India could not stake a claim to it. Besides, Bhutan was not a protectorate of India as Sikkim had been.[54] Moreover, Indo-Bhutan relations were based upon a treaty of 1949 in which Bhutan's independence was specifically acknowledged. If Bhutan foresaw any threat to its survival as a full-fledged member of the United Nations, it was free to raise its voice in international forums, and act as an independent state.[55]

As a result, since the 1970s, after becoming a member of the United Nations, Bhutan intensified its efforts to diversify its relations from India. For instance, during the coronation ceremony of the King of Bhutan in June 1974, it invited more than 150 foreign delegates. Representatives from the U.S., Britain, France, China, the Soviet Union, Canada, Switzerland, Australia and New Zealand attended this function.[56] It was the first time that Bhutan had opened its doors to the outside world. This gathering of major powers in Thimpu boosted its confidence as a sovereign power. The Royal Government of Bhutan put forward two demands before the Government of India in the same year. First Bhutan proposed to open offices in Bonn, London, New York and Paris and to issue visas to encourage and facilitate tourist travel to the country.[57] Second, it wanted to resume trade with Tibet, a practice which reportedly existed before Lhasa was taken over by Beijing.[58] These two demands were put forward by Bhutan in the light of the economic backwardness of that country, and its urgent requirement for foreign exchange for its developmental efforts.

Against this backdrop, the Government of India invited the King of Bhutan to New Delhi. He visited India on 18 December 1974, and reaffirmed strong and friendly Indo-Bhutan relations on the basis of the treaty of 1949. His visit, to quote him, was 'to reaffirm the warmth and friendship that has traditionally characterized Indo-Bhutan relations'.[59] It was his first visit abroad after his coronation. He met the President of India, F. A. Ahmad, and Prime Minister Mrs. Indira Gandhi. He assured them that he would follow the guidelines of his father to further strengthen the bonds of friendship between India and Bhutan. Both countries expressed satisfaction with the existing arrangement between them. For instance, India agreed to assist and fulfil the objective of Bhutan's third five year plan, and talks were held on the issue of the construction of the Chukha Project and a cement plant at Pagali.[60]

In August 1976, the King of Bhutan again visited New Delhi to strengthen the Indo-Bhutan relations. He appreciated the economic and developmental assistance given by India. He also deeply appreciated 'the friendship, generosity and understanding' shown by the Government of India towards Bhutan and its people.[61] Though the emergence of the new regime, led by the Janata Party Government in 1977, changed the direction of India's foreign policy and followed the 'policy of genuine non-alignment and beneficial bilateralism',[62] it did not alter the Indo-Bhutanese relations. The execution of the policy of 'beneficial bilateralism' had three important dimensions, namely personal rapport, economic accommodation, and political neutrality and non-interference in internal affairs.[63] The Janata Party Government appointed J. P. Hiremath as Indian Representative to Bhutan. While presenting his credentials to the King of Bhutan, he assured Bhutan that 'India had no interest except Bhutan's progress and the mutual security of both the countries'.[64] This assurance was significant in view of Bhutan's apprehension regarding India's intentions following the earlier merger of Sikkim in 1974. In response, the King of Bhutan expected that India would continue to assist it in every possible way towards the accomplishment of its goal of economic self-reliance.[65]

This was followed by the Indian Foreign Minister Atal Behari Vajpayee's visit to Bhutan in November 1977. Vajpayee stated that his government would 'remove the scope for misunderstading in our relations with our neighbours, avoid the build-up of any situation which would lead to irritants at the political level and thus contribute to the further consolidation of the basis of mutual trust and confidence in each other'.[66] For instance, in the past, Bhutanese students with an offer of foreign scholarships, were required to obtain a 'P' form from India for travel to those countries.[67] During this visit, the Government of India removed this irritant.

During his visit to New Delhi in March 1978, the King of Bhutan acknowledged generous Indian support and assistance to Bhutan's effort at modernization and economic self-reliance.[68] He also mentioned that Bhutan and India had developed genuine friendship and trust, and had based their relationship on common interests and aspirations. He said:

> Mr. President, I want you to know that my people and I place our complete trust and faith in our friendship with India. I am confident that both India and Bhutan will continue to do everything possible to further strengthen Indo-Bhutan friendship, which is so important to the interest of both the countries.[69]

India continued with its assistance to Bhutan, particularly in the survey of the Begana and Raidak hydroelectric schemes, and a number of micro-hydroelectric power projects in Bhutan.[70]

However, in July 1978, Bhutan expressed its concern over the interpretation of Article 2 of the treaty of 1949. In a statement, Bhutan reminded the Government of India that, 'in regard to its external relations, it would be entirely up to the Royal Government of Bhutan to decide whether to accept such advice or not. It is not correct to say that Bhutan's future still depends on Indian goodwill and friendship.'[71] These statements reflected Bhutan's eagerness to seek more and more independence from the Indo-Bhutan Treaty of 1949.

In the year 1979, during the Non-Aligned Summit at Havana, Bhutan took a position on the Kampuchean (Cambodian) issue which was not only contrary to that of India, but also favoured China. 'India wanted to keep the Kampuchean seat vacant, while Bhutan voted to allow it to be occupied by the representation of the ousted Pol Pot government.'[72] While explaining his stand on this issue, the King of Bhutan said, 'Our stand on Cambodia is a small step forward.'[73] Thus Bhutan's small step forward actually became a stepping stone in its independent foreign policy especially in the context of the Indo-Bhutan Treaty of 1949. The King asked for 'updating' the treaty with India during his brief stopover at Bombay on his way back home from the Havana Summit. He stated that, '[T]here are no problems between our countries and our friendship is deeper today than in 1949. But why leave any loose interpretation? It will be to the advantage of both India and Bhutan to update the treaty.'[74]

Two incidents that impelled Bhutan into asserting its independence of action under the Indo-Bhutan Treaty of 1949 may be cited here. One was an occasion when the Indian Foreign Minister, S. N. Misra, passed unsavoury remarks about Bhutan's support to the Kampuchean regime at the Havana Non-Aligned Summit. He said Bhutan had given vent to the feelings of some of the small countries by supporting Pol Pot at Havana.[75] The other incident was the circulation of newspaper reports to the effect that the Chinese were intruding into Bhutan creating a crisis situation there. It was further alleged that Bhutan was moving closer to China.[76] The King of Bhutan reacted strongly to this allegation and asserted that, 'it was utter nonsense to say that Bhutan was turning towards China'.[77] However, this misunderstanding was removed when the King of Bhutan clarified that Bhutan's National Assembly had already passed a unanimous resolution in 1978 not to open trade or diplomatic missions with China.[78] He said, 'though Bhutan did not want trade with China, it did not wish to annoy or provoke Peking either, in any way'.[79]

Thus one can say and observe that Bhutan's foreign policy had become both more sensitive and assertive at a time when there was an absence of a strong government in India.

In 1979, when the Janata Party Government in pursuit of its policy of 'Beneficial Bilateralism' initiated steps towards normalization of Sino-Indian relations, some members of the Bhutanese National Assembly expressed the desire to establish direct contacts with China.[80] So, Bhutan approached India in this regard, especially, 'since the Sino-Indian war of 1962, Bhutan was insisting on bilateral talks with China'.[81] In the meantime, India had surveyed Bhutan's border with the Tibet region of China and helped the country establish its own survey department in 1972. Finally, with the establishment of a Boundary Commission in 1981, the Government of India allowed Bhutan to hold bilateral talks with China. And to make further progress in that direction, Bhutan sent a diplomatic note to the Chinese Embassy in New

Delhi in March 1981.[82] Thus Bhutan established informal contacts with China.

All these developments show that Bhutan went through a gradual process of diversification in the Indo-Bhutan relationship, particularly during the period 1975–80. It was a time of political instability in India due to the declaration of the emergency, and the coming to power of the first non-Congress government. Bhutan did not lose any opportunity to assert its independent foreign policy and project its identity in world politics during this period.

Sino-Bhutan boundary issue

As mentioned earlier, Chinese occupation of Tibet in 1950 caused grave apprehension in Bhutan which were confirmed when, in July 1958, Communist China laid claim not only to vast portions of Indian territory, but also to some 200 sq. miles of Bhutanese territory as being part of Tibet.[83] Further, in July 1959, in pursuance of its policy of integrating Tibet with the heartland, China seized control of the Bhutanese-administered enclaves in western Tibet in the vicinity of Mount Kailash and the Gartok region.[84] The area of south-west Bhutan, which is strategically important due to its topographical features, 'provides an excellent observation point over the Chumbi valley and the roads leading to it'.[85] Since this area is closer to the strategic Jaldhaka barrage in the Indian state of West Bengal, China does not want to forego its claim on this disputed area. Hence, the Chinese have constructed a road linking the Chumbi valley with Bhutan.[86]

Thus, the boundary issue with China created serious problems for the Bhutanese. It became a more crucial part of its foreign policy objectives, as China did not want any Indian interference in this matter. 'In fact, Chinese armed intervention and territorial encroachment into Bhutan aimed at pressurising Bhutan to accept the idea of direct talks.'[87] Besides, the Chinese encroachments in the unpatrolled areas of Bhutan and its permanent settlement of its armed herdsmen,[88] pointed to Chinese aggressive postures.

Bhutan started direct talks with China in 1984 to sort out its boundary problems. Bhutan established informal contacts with China through a diplomatic note sent to the Chinese Embassy in New Delhi in March 1981. Bhutanese Foreign Minister, Dawa Tshering, at the Tshogudu expressed his keen desire in June 1981 to have direct talks with China. He said, 'Bhutan would like to start direct and bilateral negotiations with China to delineate and demarcate the Bhutan China boundary, and to maintain the status quo of the present Bhutan traditional border until such time as the border negotiations are concluded.'[89]

The first round of boundary talks was held at Beijing in 1984. At this time, China made it clear that its approach to the boundary issue with Bhutan would be the same as in the case of Pakistan, Burma and Nepal.[90] This meant that China recognized the independent and sovereign status of Bhutan. The second point on which China was firm was that it would not accept Bhutanese claims on strategic points. It added that it would be more generous on less important territorial claims, but not on strategic points which are more advantageous to Bhutan and India.[91] China was pressurizing Bhutan into accepting its claims.

The second round of talks was held in April 1985 at Thimpu, when both sides realized that there was not much of a dispute between the positions[92] over their traditional boundary. It was during the third round of border talks held at Beijing in June 1986, that the Chinese Premier, Jiang Zemin, assured the Bhutanese delegation that China would not interfere in Bhutan's internal affairs and the two countries vowed to maintain a 'peaceful and friendly border' while seeking an early settlement.[93]

The fourth round of border talks was held in Beijing from 10 to 14 May 1988. During these talks, both countries concurred on a four-point pattern of guiding principles which would govern their mutual relations and issued a joint statement to this effect. These guiding principles were:

- Observing the five principles of peaceful coexistence:
    - mutual respect for sovereignty and territorial integrity;
    - mutual non-aggression;
    - non-interference in each other's internal affairs;
    - equality and mutual benefit and peaceful coexistence.
- Treating each other on an equal footing and entering into friendly consultations on the basis of mutual understanding and mutual accommodation with a view to reaching a just and reasonable settlement.
- Taking account of the relevant historical background based on traditions, customs, usage and administrative jurisdiction while accommodating the national sentiments of the people and national interest of the two countries.
- Pending final settlement of the boundary question, maintaining tranquillity on the border and the status quo of the boundary as before March 1959, and refraining from unilateral action or the use of force to change the status quo of the boundary.[94]

Both sides exchanged views on the demarcation of the Bhutanese–Chinese border. The two sides reiterated their desire to maintain a peaceful and friendly border, and expressed their keenness to make every effort for an early demarcation of the boundary.[95] They were satisfied with the result of the talks. The two delegations also had a useful exchange of views on international issues of mutual interest and concern and briefed each other on the efforts of their countries towards economic development and commended the achievements made by each other.[96] In this way, the first four rounds of talks were utilized to finalize the four guiding principles.

The fifth round of border talks took place in Beijing in 1989, where substantive talks were initiated. Both countries had claimed the 495 km^2 area of Pasamlung and Jakarlung valleys as part of their respective territories. In particular, the Bhutanese delegation claimed that these valleys were Bhutan's territory on the ground that they lay below the source of the Bazaraguru Chhu river of Kurote Dzongkhang in Bhutan.[97] In response to the Bhutanese claims, the Chinese delegation offered a package proposal to Bhutan at the sixth round of talks, which took place in Thimpu in August 1990. According to this proposal, the Chinese 'offered to give Bhutan the Pasamlung and Jakarlung valley, with an area of 495 km^2 in the central sector of the boundary, if Bhutan agreed to concede the 269 km^2 disputed area in the western sector'.[98] This offer was made to Bhutan, as the 269 km^2 area in the western sector had great strategic significance to China for its security. However, the Bhutanese delegation did not agree to the Chinese proposal. Bhutan's Foreign Minister, Dawa Tsering stated that the border talks between the two countries, which were held in a cordial atmosphere, continued to make steady progress towards resolution of the boundary problem.[99] He also stated that the border issue could not be resolved overnight and that both sides needed to show patience and tolerance and make every effort to resolve the problem.[100]

On this occasion, the Chinese representative Qi Huaiyuan, Vice-Foreign Minister of PRC, said that China and Bhutan had many similar views on international issues.[101] Interestingly the Bhutanese Foreign Minister said:

> Bhutan shared China's aspiration for peace and stability in Asia and the world to provide a better and more dignified life for its people. Bhutan respected China's adherence to the five principles of peaceful coexistence, which was its basis for relations with other states.[102]

On the occasion of the 74th session of the National Assembly in 1996, the King of Bhutan said that after the last round of boundary talks between Bhutan and China, which began in 1984 when there were more than 1,000 km^2 of territory under dispute, there are essentially only three areas in the western sector, which are still under negotiation. These are 89 km^2 in Doklam, 42 km^2 in Sinchulumpa, and 138 km^2 in Shakhatoe – a total of 269 km^2.[103] The king described the Bhutan–China boundary as an important issue as it affected Bhutan's national interest and the wellbeing of the Bhutanese people.

The tenth round of talks concluded in Beijing in November 1996. Its Foreign Minister and the Secretary of Survey represented Bhutan. They put forward Bhutan's claims to Doklam, Sinchulung, Dramana and Shakhatoe in the western sector of the northern border and stressed that these were vital as pasture lands for the people of the Haa valley.[104] The Bhutanese delegation also informed their Chinese counterpart that Tibetan herdsmen had been intruding into Majathang and Jakarlung in the central sector of the boundary and had even constructed sheds.[105]

According to the Foreign Minister of Bhutan, 'Chinese Officials did not make any comment on the Bhutanese territorial claims but on the question of intrusion by Tibetan herdsmen, they pointed out that since there was no agreement on the proposal (offered by the Chinese during the seventh round of talks), they could not control the activities of Tibetan herders along the borders'.[106]

The eleventh round of border talks was held in Bejing in 1997 in which China proposed to exchange an area of 495 km² of Pasamlung and Jakarlung valleys in the northern borders of central Bhutan for Sinchulump, Dramana and Shakhtoe being an area of 260 km² in northwest Bhutan. In addition to this, China proposed to sign an interim agreement for the maintenance of peace and tranquility along the borders. However, this agreement was finally signed during the twelfth round of border talks on 8 December 1998.[107]

The twelfth round of bilateral border talks was held in Beijing from 8 to 12 December 1998 and Jigme Thinley, the Head of the Cabinet, represented Bhutan. The talks focused on three important points:

- the establishment of diplomatic relations between Bhutan and China;
- the establishment of trade relations with China; and
- the question of the exchange of land.[108]

Bhutan and China also signed an 'Agreement to Maintain Peace and Tranquillity' on the Bhutan–China border in 1998, which is the first Sino-Bhutanese agreement or treaty at present. There was no final decision on the exchange of territories as the proposed area to be exchanged had borders with the Indian state of Sikkim. It showed Bhutan's concern not to jeopardize India's interest in the Himalayan region while maintaining its formal relations with China.

The thirteenth round of border talks took place in Thimpu in September 1999. China was represented by Wang Yi, Assistant Foreign Minister. He stated that he had come to Thimpu for 'signing the last year's intergovernmental agreement between the two countries, which has special importance and significance'. On the issue of friendship between the two countries, he said:

> China has always pursued a policy of developing friendly relations with all its neighbouring countries on the basis of five principles of peaceful coexistence . . . and we hold the view that countries, big or small, rich or poor, strong or weak, should all be equal members of the international community. We have, all along, treated Bhutan as an equal neighbour.[109]

The fourteenth round of boundary talks was held in Beijing in November 2000. Bhutanese Foreign Minister Jigme Thinley led a seven-member delegation to China and met Chinese Premier Zhu Rongji and Chinese Foreign Minister Tang Jiaxuan. Both countries 'expressed their satisfaction with the progress being made on the boundary discussions and agreed to continue the dialogue in accordance with the "Five Principles of Peaceful Co-existence" '.

The Sino-Bhutan relationship thus developed steadily since the two countries had signed an agreement on the maintenance of peace and tranquillity in the border area in 1998.

During this round of talks, Bhutan extended its claim to the line of the border beyond what the Chinese Government had offered. Bhutan also proposed technical discussions between experts from the two sides using maps. The king told the National Assembly of Bhutan on 14 July 2001 that the proposed extension of the border along the three sectors under discussion was in the Doglam, Sinchulumba and Dramana areas. He expressed confidence that Bhutan 'being a small country' and China 'being a very large and friendly neighbour', the demarcation of the boundary would be completed successfully in the near future.[110]

The fifteenth round of boundary talks was held in Thimphu, in November 2001. Vice Foreign Minister Wang Yi led the Chinese delegation whereas Foreign Minister Lyonpo Jigme Yoeser Thinley represented Bhutan. Since the discussion was focused on the international boundary between the two countries, Wang Yi said:

> 'As China and Bhutan are neighbouring countries, with shared mountains and rivers, we should first of all achieve mutual understanding and, on the basis of mutual understanding, we can achieve mutual support and mutual confidence and mutual assistance. We have made a lot of achievements in this regard. We will work on this basis to make further progress.' He added 'it is my view that what counts most is that we must continue the process and, during this process, enhance the mutual understanding so we can work together to remove the outstanding relatively minor issues.'[111]

The sixteenth round of boundary talks took place in Beijing, China on 12 October 2002. Bhutanese Foreign Minister Jigme Thinley met the Chinese Premier Zhu Rongji, who stated that 'China and Bhutan are resolving their boundary issues through mutual understanding and reconciliation' and the sixteen rounds of talks to date had achieved successes.[112] He expressed China's appreciation of the Bhutanese Government's support on issues like Taiwan and human rights, and hoped to develop co-operative relations based on the principle of equality, mutual benefit and non-interference in each other's internal affairs.[113] In response, Bhutan's Foreign Minister reiterated Bhutan's consistent stand on the issues of Taiwan and human rights and emphasized that Bhutan would continue to firmly support China on the issues.[114]

After so many rounds of talks between Bhutan and China, one can surmise that the Chinese have been reluctant to reduce their claims on the western sector, which is strategically important for them, or to interfere in the question of Tibetan yak herders' intrusion into Bhutanese territory, unless and until the Bhutanese accept the proposal of exchange of an area of 495 km^2 with the pasture lands covering an area of 269 km^2. On the other hand, Bhutan was reluctant to accept the Chinese proposal, as a large number of Bhutanese people depend on livestock for their livelihood, these pasture lands being vital for their survival. The Bhutanese also claim that both the areas had always belonged to them.[115] Bhutan wants to ensure that its traditional grazing land is not affected and that the boundary is demarcated according to traditional lines.[116]

Until April 2004 seventeen rounds of talks had been completed but there was no permanent resolution to this issue. This delay in any concrete decision by both parties created a lot of misunderstanding in Indo-Bhutan relations, since India's strategic interest is very much involved in the Himalayan kingdom. On the issue of further delay in final resolution of boundary issues, Bhutan made a typical diplomatic answer: 'Big objectives take time to fulfil.'[117] The Foreign Minister of Bhutan expressed optimism about the final outcome of the boundary talks. He added, 'it would be better to conduct the talks without haste and with great care and patience as it involves the national interest of the country'.[118] The Foreign Minister said that a small country like Bhutan must be tactful in its approach to its boundary negotiations with its large neighbour because it cannot afford to lose even a single square kilometre of land.

To resolve the demarcation of the boundary line with China and also the question of intrusion of Tibetans in Bhutanese territory, and while doing so, maintaining good relations and an atmosphere of friendship and co-operation with China, are the main objectives of Bhutan's foreign policy.

Making an analysis of the overall Sino-Bhutanese relations, one can see its serious implications for India–Bhutan relations. From 1958 to 1962, China's military might always threatened Bhutan's territorial integrity, peace and tranquillity of the Himalayan region. India's defeat in the 1962 war raised many doubts in the minds of Bhutan's ruling elite, who voiced concern against too much dependence on India. At this juncture, some officials in Bhutan insisted that Bhutan should have developed closer relations with China to counter-balance 'India's neo-colonial political and economic presence in the country'.[119]

However, since the mid-1980s, the beginning of direct boundary talks with China, China consistently offered to establish economic and diplomatic relations with Bhutan, which has implications for the India–Bhutan Treaty of 1949. Although Bhutan is not interested in using China as a counterbalance for Indo-Bhutan relations, it is more inclined to maintain reasonable relations with it without offending India's interest in the region. To quote Leo E. Rose, 'no doubt, the Royal Government wants China as a friendly (or at least non-threatening) neighbour, but one with whom relations are correct rather than intimate'.[120]

This is how the foreign policy of a small state like Bhutan works for its survival when it is sandwiched between two powerful countries. Such a policy cannot but be ambivalent and non-antagonistic. A small state has little choice other than to maintain a low profile in dealing with its larger and more powerful neighbours. This is the strategic stand taken by Bhutan vis-à-vis China.

Bhutan's ethnic crisis and pro-democracy movement

In the 1990s, Indo-Bhutan relations were characterized by a sort of give-and-take relationship. The King of Bhutan paid regular visits to India and praised Indian efforts for the economic development of Bhutan. He appreciated the steady relationship, which was based on mutual trust and equality. In return, the king expected a positive attitude from India on the issue of Nepali immigrants. For instance, in July 1990, when the Indian Energy and Civil Aviation Minister Mohammed Arif Khan visited Bhutan, the King expressed his full confidence and complete satisfaction with the Indo-Bhutan relationship.[121] In the same year, Bhutan invited Jyoti Basu, Chief Minister of the neighbouring state of West Bengal, to discuss the issue of Nepali immigrants. In response, Basu assured Bhutan that he would not permit the territory of West Bengal to be used for any agitation against Bhutan and decided to deploy police forces along the border to prevent any such activities.[122] The Government of India took the same position. The Indian Foreign Secretary, Muchkund Dubey, claimed that the Indian Government had already warned the Nepalese settlers in India that it would not tolerate their anti-Bhutan activity.[123] Despite these assurances from India, the Indo-Bhutan relationship came under severe strain particularly due to the growth of the pro-democracy movement in Bhutan.

The disproportionate presence of ethnic Nepalis in Bhutan resulted in a sense of apprehension among the ruling *Drukpas* of their being outnumbered as had happened in Sikkim. The concept of a 'Greater Nepal' and the cohesiveness of the Nepalese migrants, who had become the majority community in Sikkim as well as in the districts of Darjeeling and Kalimpong and parts of north-east India, added to the fears of ethnic *Drukpa* Bhutanese. These fears compelled Thimpu to enforce immigration laws vigorously and force those being unable to produce sufficient proof of their domicile prior to 1958, to leave the country. This policy resulted in widespread resentment among the southern Bhutanese, who felt they were being made aliens in their own land due to their ethnic identity. Now Bhutanese of Nepali origin reorganized their movement to bring a change in the political system from a monarchy to a democracy and campaigned all over the country against the ill-treatment meted out against them. To control these activities, the Government decided to bring the Bhutanese of Nepali origin into the mainstream *Drupka* culture. The King of Bhutan introduced an ideology of 'One Nation, One People' which led to the imposition of a code of conduct, including *Drukpa* language, religion and dress for all Bhutanese citizens through a policy of *Driglaham Namzha* (a revival of traditional Bhutanese culture). The main objective of this policy has been to establish national homogeneity and cohesion among the various linguistic and ethnic groups in the country. In addition to the *Driglaham Namzha* policy, the government adopted various policies and passed several resolutions in its National Assembly that went against the interests of the Bhutanese of Nepalese origin. For example, in 1991 it was decreed that anyone involved in peaceful demonstrations or suspected of supporting human rights movements would be evicted. Forcible eviction was legitimized with the introduction of voluntary migration forms that were printed in the *Dzonkha* language, which most Bhutanese of Nepalese origin cannot read. As a result, many who filled in the 'voluntary migration form' to the best of their knowledge, indirectly surrendered their citizenship of Bhutan. This resulted in large number of people becoming refugees in Bhutan.

In an attempt to defuse the situation, bilateral level talks between Bhutan and Nepal began in 1990. Eventually, both countries agreed to classify the refugees into four categories:

- Bhutanese who had been forcefully evicted;
- Bhutanese who had voluntarily emigrated;
- non-Bhutanese;
- Bhutanese who had committed criminal acts.

Out of these four categories, Thimpu agreed to the return of people belonging to the first category, although it felt that most of the refugees in United Nations High Commission for Refugees camps in eastern Nepal were originally Nepalese and Indian citizens looking for free food and shelter and health care.[124] However, in subsequent bilateral talks, particularly during 1993–96, both Bhutan and Nepal took more rigid stands regarding the repatriation of Bhutanese refugees. Further progress in the bilateral talks stalled due to the ongoing political instability in Nepal, with each of its new administrations changing the country's stand on the issue. During the eighth round of bilateral talks held in September 1999, Thimpu acceded to international pressure, particularly from funding agencies and human rights organizations, by agreeing to take back the refugees belonging to the second category.[125] During the ninth and tenth rounds held in May 2000 and in December 2001 respectively, both countries finalized the details to start the process of verification of the people in the camps.

As a result of the process of rehabilitation of Bhutanese refugees, the details of verification and categorization of the Khundunabari camp was released unofficially. It has been disclosed that out of the total camp population of 12,183 some 293 (2.5 per cent) were adjudicated to be bona fide Bhutanese. It means that they had acceptable proof (such as identity documents, property deeds or tax certificates) of being Bhutanese. The vast majority of the refugees comprised the Bhutanese who had emigrated voluntarily.[126]

Unfortunately, the joint statement did not mention what repatriation procedure would be followed and no particular date was set for beginning and ending this process. However, by the end of 2003, under pressure from international donors, most of them being European countries and the UN High Commissioner for Refugees (UNHCR), both Bhutan and Nepal finally agreed to conduct bilateral talks on the refugee issue. Accordingly, the 15th Ministerial Joint Committee Meeting was held in Thimpu from 20 to 23 October 2003,[127] and the repatriation of those verified was to begin by the end of February 2004. But in reality, the repatriation process could not take place due to an 'untoward incident'.[128] Since then, there has been no progress in resuming the repatriation process or any sign of further steps towards new methods of verification of these refugees.

As far as India is concerned, traditionally it has had close relations with both Bhutan and Nepal. Thus, despite having the potential of becoming an independent arbitrator, India is finding itself in a dilemma over supporting the cause of Bhutanese refugees. Moreover, India has been very much involved with its own problems of insurgency on its Himalayan frontiers, Maoist insurgency at the Indo-Nepal borders and challenges from north-east Indian insurgents at the Indo-Bhutan border. India also supported Bhutan's 'Operation All Clear' launched by Bhutan's Royal Army. Thus India could not to take any initiative in this regard and followed a 'wait and see' policy.

From mid-2004 to mid-2006, there was no positive sign of resolving this refugee crisis on account of the process of democratization in Bhutan and the continuing political instability in Nepal because of challenges from Maoist insurgents and suppression of civil and political rights in Nepal by the monarchy. In October 2006, the United States made an official announcement expressing its willingness to accept 60,000 Bhutanese refugees living in Nepal. The UNHCR described this offer as an opportunity to break a long-standing deadlock over more than 100,000 refugees from Bhutan living in eastern Nepal. 'Years of bilateral negotiations between Nepal and Bhutan have made little progress in resolving this issue,' said UNHCR spokesperson Jennifer Pagonis at a briefing in Geneva on 6 October 2006, 'and the opportunity of large-scale resettlement is a real spark of hope'.[129] The US Assistant Secretary of State for Population, Refugees and Migration, Ellen Sauerbrey, made the resettlement offer while attending the UNHCR's Executive Committee meeting in Geneva from 2 to 6 October 2006. After 15 different consultations between the two Governments', Sauerbrey said at a briefing on 2 October 2006, 'the United States has come forward and said we are willing to resettle a very significant part of this population'.[130] She also stated that it is likely that the United States will be able to absorb up to 60,000 refugees over three or four years, with Canada and Australia also offering to take some of them.[131]

But it seems that the US offer of resettlement did not attract Bhutanese refugees and most of the leaders from the refugee camps have seen this move as creating a division amongst them. They were actually in favour of being repatriated to their home country.[132] The Chairman of the National Front for Democracy of Bhutan, Thinley Penjore stated:

> It is unfortunate that America's offer to accept 60,000 refugees has not come officially through the Nepal Government and it has created confusion in the camps and has created a rift among the refugees. Had it come through the Government, it would have been brought to the notice of the refugee leaders before going to the people's level directly.

The UNHCR Representative to Nepal, Abraham, said that the issue of resettlement of refugees in the third-party countries was only an option. He said it was up to the refugees to decide whether or not they are prepared to take up the offer. He stated that in a democratic set-up, the refugees cannot have imposed on them any decision, either that of the UNHCR or the international community.[133] Thus the overall reaction towards the offer from third-party countries to resolve the crisis is mixed: some of the refugees are interested in being resettled in the US, Canada or Australia or any of the EU nations. And those who are politically motivated and fighting for the democracy in Bhutan are not willing to abandon their hope of returning home.

The Government of Nepal took the stand that it will give first priority to resolving the crisis at the bilateral level, and then it will choose an option to bring India into the picture or globalize the issue. To quote Nepal's Deputy Prime Minister and Minister for Foreign Affairs, K. P. Oli, 'his talks with Bhutan would be finalized at the bilateral front. If it did not resolve the problem, we will go for the second option and then the final option of globalizing the issue.'[134]

On its part Bhutan welcomed this offer and agreed to co-operate with the UNHCR Team for surveying the genuine refugees from the camps. In its editorial, *The Kuensel* (the mouthpiece of the Royal Government of Bhutan), stated:

> The international community has commended the offer that is backed by proposals from the Australian and Canadian governments, which have also offered to take an unspecified number of people to their countries. The world has acknowledged that there is a viable solution to a humanitarian problem that looked like it could drag on indefinitely.[135]

Ironically, India did not come forward with any official statement or policy decision about the refugee crisis. India's passive stand on the refugee crisis is indirectly favouring Bhutan, leaving scope for it not to take any concrete steps towards the final solution while maintaining close and cordial relations with India. Moreover, Bhutan has been assured by the Government of India that it would extend all possible help to it. India is treating this problem as Bhutan's internal affair, which should be solved through bilateral talks between Bhutan and Nepal.

North-east militant camps in Bhutan

It is a well-known fact that many militants from Assam, especially ULFA (United Liberation Front of Assam) and NFDB (National Democratic Front of Bodoland) had taken shelter in Bhutan after a major Indian military offensive at the beginning of 1990. Now these militants have multiplied into a few thousand cadres working for various organizations like BLTF (Bodo Liberation Tigers Force) and KLO (Kamatpuri Liberation Organization), as the geo-strategic location and mountainous terrain of Bhutan provides them safe sanctuary. These groups are fighting against the Government of India as well as the Government of Assam and its police force. To meet the secessionist challenges from the Assamese militants and their hideouts in the Indo-Bhutan border, particularly the areas bordering Assam and Bhutan, the Indian Government proposed to initiate a joint Indo-Bhutan army operation to drive out these militants. In the 78th Session of Bhutan's National Assembly, Tshongdu, held in June 2000, it passed a resolution for a four-point plan to flush out insurgents from Bhutanese territories. The resolution laid down the Bhutanese action plan:

- to cut off ration supplies to the militants;
- to punish all groups and individuals found helping the militants by invoking the National Security Act;
- to pursue the process of dialogue with the militants to make them leave peacefully; and, if all efforts fail,
- to take military action as a last resort.[136]

The Royal Government of Bhutan applied pressure on various Indian terrorist groups, including ULFA, NDFB and a third group, Kamtapuri Liberation Organization, to remove their camps from Bhutanese territory. Under an agreement between ULFA and the the Royal Government on 18 June 2001, the former had agreed to shut down four of the nine camps they had been operating in that country, and to relocate their cadres to some other destination, by 31 December 2001. Although the ULFA had closed down the four camps – one each in Gobarkonda, Nangri, Deori and the Military Training Centre in Martshala – it simultaneously opened a new camp on a mountain ridge above the Samdrup–Jongkhar–Trashigang highway.[137]

There were an estimated 4,000 ULFA and about 1,000 NDFB militants, as well as an unspecified number of KLO cadres holed up in Bhutan. According to Bhutan, ULFA had six known camps; the NDFB had seven, consisting of three main camps and four mobile camps between Lhamoizingkha and Daifam, and the KLO had two known camps. However, certain unconfirmed Indian reports indicated that ULFA, along with the NDFB, ran more than 30 training camps in Bhutan. It was also reported that several training camps were jointly run by the ULFA and KLO. The terrorists based in Bhutan have carried out a series of attacks on security forces in Assam and West Bengal.

They also targeted civilians and vital installations in India. After attacking their targets in India, these terrorists invariably return to their bases inside Bhutan.

According to various intelligence and security agencies, many Bhutanese Government officials were involved in supporting the ULFA militants through economic and military aid. To quote Jaideep Saikia:

> A certain Bhutanese Army Brigadier, V. Namgel, Security-in-Charge and Military Adviser to Bhutanese King, has been actively helping the ULFA in obtaining arms and ammunition from foreign countries.[138]

There were many other incidents which indicated the involvement of Bhutanese personnel in supporting ULFA militants. For instance, two staff members of the protocol division under Foreign Ministry Wangchuk Dorji and Lhaba Tshering, were caught by its own police while transferring US$38,000 and Rs.300,000 in diplomatic pouches.[139] Responding to this, the Bhutanese Government denied all these reports and stated that the two officials had 'acted on an individual level and their actions had nothing to do with the Government'.[140]

To make matters worse, the ULFA militants, while seeking sanctuary in Bhutan, have been targeting the innocent Bhutanese civilians and businessmen on the Assam–Bhutan border. Moreover, ULFA leaders were also demanding a soft policy from Bhutan towards them. To quote ULFA Chairman, Arabinda Rajkhowa:

> We reiterate that ULFA is not going to occupy Bhutan permanently, nor is there any plan to include Bhutan in a sovereign Assam. Our camps in Bhutan are not for such purposes. As a result of the Indo-Assam conflict, our freedom fighters of Assam had to establish camps in Bhutan. It is a basic human right to be able to go to a place even when one's life is in danger . . . We appeal to the Government of Bhutan and the Bhutanese people again that the cadres of ULFA be given the universal right to maintain the centuries old Assam–Bhutan cordial relationship intact.[141]

In the meantime, the Bhutan Government made several attempts to resolve this crisis in a peaceful manner. In March 2003, it reportedly asked the terrorist groups to vacate their camps from Bhutan by 30 June 2003. However, the groups did not respond to this deadline and continued their violent activities on the border by killing innocent civilians. Bhutan realized the gravity of the situation and during the four day visit of the king to India in September 2003, Bhutan and India shared their common security concerns over the presence of nine ULFA camps, eight NDFB camps and four KLO camps in Bhutan.[142]

Addressing a formal banquet hosted by the President of India, the King of Bhutan said:

> The continued presence of these armed militant groups undermines Bhutan's security, while their goals and activities in Assam and West Bengal are a threat to India's peace and territorial integrity. Our two Governments are in close touch regarding this serious problem and Bhutan is fully committed to the understanding between us as close friends and allies that we will not allow our territories to be used by anyone for carrying out activities that are harmful to each other's national interests.[143]

In addition to this, he said 'the National Assembly had already taken a decision to make a last attempt to invite the leaders of the ULFA, NDFB and KLO for talks to resolve the crisis through a process of dialogue'.[144]

Whereas Bhutan did not want to use armed force to tackle this problem, it also wanted to avoid any antagonism from India. It feared that the use of armed force might lead to an unholy nexus between the militants and the suppressed anti-monarchical forces within and outside Bhutan. Therefore, it was in the interests of Bhutan's security to continue a peaceful dialogue with the ULFA militants while taking the Indian Government into its confidence. Bhutan informed India about the impending action against the Indian insurgents on 13 December 2003.[145] Accordingly, the 'Prime Minister conveyed to His Majesty the King that the Government and people of India would stand firmly and solidly behind the Royal Government of Bhutan at this critical juncture and would provide all the necessary support as requested, till the task is completed'.[146]

In the meantime, Bhutan deployed nearly 600 militias along with regular soldiers to fight Indian separatists in the five southern districts of the Kingdom where the three groups (ULFA, NFDB and KLO) had at least 22 camps including the military training bases.[147] According to Lt. Col. Dorji Khandu, 'a total of 591 trained militia volunteers, including 20 women, were sent to the southern districts as part of a training and familiarization programme'.[148] Thus 'Operation Flush Out' began in early December 2003 and was successful in its mission of either chasing the Indian militants up to the border or killing them. It encouraged the Bhutanese Army to initiate another operation known as 'Operation All Clear' so that there would be no militants left in hide-outs on Bhutanese soil. On 13 December 2003 the Bhutan Government issued a final notice to the insurgents to quit Bhutan and by 15 December 2003, it launched 'Operation All Clear'. On its part, India promised Bhutan that the Indian army would be taking necessary measures to intercept movement of militants from Bhutan to India. The state Governments of Assam and West Bengal were alerted to deal with the situation arising out of this action, including sealing the borders and maintaining the peace.[149] In return, Bhutan assured the Government of India that it would not allow its territory to be used for activities inimical to India's interests. The launch of operations against Indian insurgent groups in Bhutan struck a blow against terrorist activities in the entire region.[150]

As a corollary to this agreement, on 15 January 2004, the Bhutan Government announced that its military offensive to expel the insurgents had been completed and the soldiers were now clearing the jungles to recover weapons left behind by the terrorists.[151] While appreciating Bhutan's efforts to flush out the insurgents from its soil, India was assured that Bhutanese civilians moving through Assam would get security against retaliatory attacks by the ULFA.[152] Tshongdu, the National Assembly of Bhutan, taking serious note of the insurgents activities, passed a resolution to construct the wall in the bordering areas with India.[153] A series of meetings was conducted by India and Bhutan on Border Management and Security to discuss the security concerns of both countries, including the Gelephu bomb blast on 5 September 2004.[154] The threat perceptions of the two countries included the dangers posed by the possible nexus between Maoists in Nepal and the people in the refugee camps in Nepal as well as between Maoist elements from the camps in Nepal and Indian insurgents in West Bengal and Assam.

In the meantime, the ULFA received another jolt when one of the founder-members of the outfit, Bhimakanta Buragohain, popularly called 'Mama' (uncle), was presented before the media along with three others by Lt. Gen. Mohinder Singh, GOC, 4 Corps, at Tezpur on 25 December 2003. In its propaganda, the ULFA alleged that 'Mama' had been captured and killed by the Bhutanese army during military operations. ULFA also demanded that the body of the dead 'Mama' should be delivered to his relatives.[155] Since there were no confirmed reports of his death or captivity by the Bhutanese army, it became difficult for India and Bhutan to cope with the situation. As a result, the ULFA was able to generate considerable public sympathy for its cause. But the Bhutanese army captured 'Mama' along with three accomplices on its soil and delivered them to the Indian army, thus exposing the ULFA's false propaganda. Apart from this, the JIBA (Joint Indo-Bhutan Army) operation compelled the senior leaders of the KLO to surrender to the West Bengal police who had fled their camps in Bhutan as a result of the military offensive.

Thus, due to the strong bonds of friendship between India and Bhutan, the insurgents from north-east India, particularly from Assam and West Bengal, realized that there was no other way than to start peace negotiations with New Delhi. Ironically, the leaders from the ULFA repeatedly requested the Bhutanese Government to act as a mediator between them and the Indian Government and to convince New Delhi to agree to their main demands.[156]

During the fourth round of the Border District Coordination Meeting (BDCM) between the Governments of Bhutan and Assam held on 24 November 2006, Tsering Wangda, the Joint Secretary of the Ministry of Home (Internal) Affairs of the Bhutan Government reassured the Assam Government, 'we will not allow any Indian insurgents to have camps inside Bhutan'.[157] He also expressed concern over insurgency in north-east India affecting the economies of Bhutan as well as Assam, which shares a border with the tiny Buddhist Kingdom.[158] Bhutan has realized the futility of providing shelter to the insurgents or having a soft corner for them vis-à-vis India. It has no other option but to maintain the status quo in its relationship with India while bargaining for its economic development.

Indo-Bhutan joint hydroelectric projects

India's enhanced economic assistance should also be seen in the context of Bhutan's internal stability which has been put under stress by the influx of ULFA militants from Assam. This has implications for India's security as well. It is, therefore, in India's interest to help Bhutan build up its economy and infrastructure in a manner that would enable it to meet its development and defence needs adequately. A brief review of various development programmes, for instance the hydroelectricity power projects undertaken by Bhutan and India, becomes necessary.

Bhutan's hydroelectricity power potential is estimated at about 30,000 MW, out of which the potential safe and exploitable resources are estimated at 6,000 MW. A very small portion of this enormous hydropower potential has so far been tapped. Accordingly India has from the beginning helped Bhutan in the development of its hydropower resources. The Chhukha Hydro Power Project is one of the biggest and the most successful projects undertaken in Bhutan and has been a symbol of Indo-Bhutan friendship and co-operation.

As part of the Royal Government's reform agenda, the power sector in Bhutan was restructured through the Electricity Act, 2001. The former Department of Power of the Ministry of Trade and Industry (MTI) was divided into three entities:

- the Department of Energy responsible for planning and policy in the power sector;
- the Bhutan Electricity Authority as an independent entity to license and regulate power companies; and
- the Bhutan Power Corporation as a transmission, distribution and supply company.[159]

The number of power connections in the country is about 52,000 (40 per cent overall access). The expansion plan sets a target of 56 per cent access by 2007 and universal access by 2020. The expansion plans are being financed by grants and soft loans from ADB and the Government of Austria. At present, the installed capacity in the country (468 MW) exceeds the demand (105 MW), and the surplus energy, about 300 MW, is exported to India.

Hydroelectricity export has become a significant source of revenue for Bhutan. For instance, in 2002, the Chhukha Hydro Power Corporation (CHPC) sold Nu.2,171 million worth of electricity to India, and the Kurichhu power project sold 97 per cent of its electricity to India.[160] The construction of the Kurichhu project started in 1995 and was completed in 2001. It was commissioned in September 2001 and began commercial operations a year later. The 60 MW project was constructed at a cost Nu.5,600 million, funded by the Government of India. According to the project officials, total domestic consumption in Bhutan is about 105 MW[161] while the rest is exported to India at a cost of Nu.1.75 per unit.[162] In 2005, the Kurichhu Project sold power worth Nu.547.2 million.

On 26 April 2006, Sudhir Vyas, the Indian ambassador to Bhutan, formally inaugurated the Kurichhu Project and appreciated 'Bhutan's people-oriented development philosophy'. He stated that this project reflected the state of wonderful relations between the people and Governments of Bhutan and India. The main objective of this project was to provide electricity to the people of eastern Bhutan and to bring about balanced regional development and prevent rural to urban migration. In fact, the people from eastern Bhutan viewed this project not as a hydropower project, but as a harbinger of economic development in the east.[163]

Another major hydropower project is the Tala Project. Located in Tala village, 60 km up from Phuntsholing, close to the Indian border, the project has an installed capacity of 1,020 MW. Started in 1996, the project was originally planned to be commissioned in June 2005. But it got delayed due to problems in excavation of tunnels and constructions of roads. The project is expected to transform the country's economic development. As a result of this project, Bhutan's annual per capita income should increase from about $700 to about $1,200. It is expected to generate revenue of Nu.40 million a day after completion.[164] As it is the biggest power project in Bhutan, and due to its proximity to India, most of the surplus power will be exported to the eastern and northern states of India. According to the Director General of Bhutan's Energy Department, almost 80 per cent of Tala's generation capacity would be traded with India. Incidentally, the Punjab State Electricity Board (PSEB) Chairman Y. S. Ratra has already placed before the Indian Government its intention to purchase 1,000 MW from the Tala Power Project.[165] It is also envisioned that the THPA operations should raise the share of hydropower dividend revenues to the total budget revenues from the current 45 per cent to 60 per cent.[166] With the help of this project by 2007, Bhutan would enjoy over-all surplus in its exports as against the present deficit in its current account. Thus, the Tala project when commissioned will bring unprecedented economic prosperity to Bhutan. In addition to this, three more Memoranda of Understanding (MoU) have been signed between Bhutan and India for the preparation of detailed project reports on the 870 MW Punatsangchu, the 992 MW Puantsangchu (stage II) and the 670 MW Mangdechu hydroelectric projects.[167] India has expressed its readiness to finance these projects as it needs power for meeting its rising energy demand.

By the year 2006, Bhutan started exporting about 6,400 MUs (mega-units) of power annually. The revenue from hydropower projects along with earnings from the traditional revenue sources could reach about Nu.15 billion annually. By 2005, this industry already contributed about 60 per cent of the total annual revenue. Thus the hydroelectricity power sector in Bhutan is the largest single revenue earner for the country's development and for upgrading the standard of living of the people.

Looking at Bhutan's land-locked situation, its low level of industrial development, lack of entrepreneurial capacity, shortage of manpower, inadequate infrastructure and the very small size of the market, Bhutan needed for its development a good co-operative and economically more developed neighbour. It found one in India. At the same time, Bhutan naturally wanted to develop a maximum degree of self-reliance, and India has assured Bhutan of its support. The Indian Government's position on this issue is reflected in Bhutan's weekly *The Kuensel*, dated 17 December 1994: 'whatever assistance we give should have in-built such provision and schemes that meet the principal aim of developing self-reliance in Bhutan'. The geographical proximity of India provides a transit route to land-locked Bhutan. It is, therefore, not surprising that most of the industries including hydroelectric projects in Bhutan are located in its strategic areas and in particular, areas bordering India, which can reduce transport costs as well as provide easy access to India.

Conclusion

An analysis of the internal and external affairs of Bhutan and their impact on Indo-Bhutan relations, shows that there are no flaws in the relationship. On the contrary, Bhutan's National Assembly has recognized India's contribution to peace and stability in South Asia. It is important here to quote what *The Kuensel* says about the Indo-Bhutan Treaty of 1949:

> But in whatever way Bhutan interprets the treaty of 1949, one thing is clear, Indo-Bhutan relations must be very smooth because India is a big neighbour and much of Bhutan's internal economy and political stability depends on India.[168]

Bhutan's policy towards India is guided by several factors such as:

- frequent Chinese incursions into Bhutan and the need for India's countervailing power;
- the importance of India for substantial economic and technical assistance to Bhutan; and
- the general realization by land-locked countries of their geographical dependence on neighbouring power(s) despite their desire for greater autonomy of action.

Bhutan continues to be guided by India, not only by the 1949 treaty, but also because of its close and centuries-old relationship with India. It has succeeded to some extent in asserting its independent status, and in liberalizing the interpretation of the terms of the 1949 treaty. In spite of some differences between India and Bhutan, there is an underlying element of friendship, and a willingness to carry out a mutually-beneficial relationship between them. This in turn has been strengthened by various geo-political, economic and strategic compulsions for both countries.

12 Security of the north-east Himalayan frontiers

Challenges and responses

Bibhuti Bhusan Nandy

Introduction

The concept of a distinct multinational Himalayan region comprising Bangladesh, Bhutan, Nepal, north-east India and north-west Burma has been gaining currency lately. Geographical contiguity of the countries in the region, its ethnic mix and purportedly shared cultural heritage, economic complementarity and foreign policy desiderata prompted the World Bank to propose formation of a transnational 'growth triangle' for the area. Irrespective of the validity or otherwise of the underlying assumptions behind such a formulation, the objective conditions in the region may pose serious obstacles to effective co-operation among the constituent countries.

The high Himalayan mountain ranges have always been an impregnable natural defence for India but the Chinese invasion in 1962 was a shocking exposé of the vulnerability of India's northern and north-eastern frontiers. Open and porous borders with Bangladesh (4,096 km) and Burma (1,859 km) have helped cross-border (often state-sponsored) low-intensity border conflicts, illegal immigration, smuggling, gunrunning and drug trafficking in the region. Poor border management resulting from the inadequacy of border guards and effective surveillance facilities, rampant corruption and political patronage of illegal movements of goods and people has greatly compounded the threats to India's internal security.

The Indo-Bangladesh land borders are mostly settled except for a 6.5 km stretch. Both India and Bangladesh claim the ownership of the New Moore Island in the Bay of Bengal. The unsettled Sino-Indian border dispute was at the root of the Chinese invasion in 1962 when the People's Liberation Army (PLA) reached as far as Haflong. The Chinese claim over Arunachal Pradesh has remained a contentious issue and incursions by Chinese border patrols into Indian territory across the Line of Actual Control (LAC) are not infrequent, at times leading to provocative situations including localized clashes. China has now restored trade between Tibet and India through Sikkim which implicitly amounts to its recognition as a state of India.

Being a buffer between Bangladesh and Burma, the access to and from north-east India is facilitated as much by artificial and porous borders as by shared ethnicity and linguistic affinities among the border populations. Consequently, clandestine trans-border movements of people and goods including contraband occur on an enormous scale.

Bangladesh threat perception

Bounded on three sides by Indian territory, Bangladesh suffers from an acutely India-centric threat perception. India's successful military intervention against Pakistan in 1971 is paradoxically a major factor in the shaping of Dhaka's threat perception. All political parties drag India into their rhetoric to score political mileage against one another. In the war-game exercises of the Bangladeshi army, India invariably figures as the target enemy stereotype. The pro-Pakistan, fundamentalist sections of the Bangladeshi media daily broadcast massive anti-India propaganda, holding New Delhi responsible for all their country's woes.

In the post-Mujib era, the military regimes of Gen. Ziaur Rehman and Gen. Ershad desecularized the country's constitution in 1977, legitimized the pro-Pakistan communal and fundamentalist Islamic parties by lifting the ban on them, and in 1988 declared Islam the state religion. As a corollary to this policy shift, Dhaka tilted towards Pakistan, Saudi Arabia and other Muslim countries, and also China as possible counterweights to India. A liberal flow of petro-dollars from West Asia led to the mushrooming of mosques and *madrassas* in every nook and cranny of the country, which has boosted the growth and spread of Islamic fundamentalism. In the 1980s participation by over 5,000 Bangladeshi *madrassa* alumni as *Mujahideen* in the *Jihad* against the Soviet occupation army in Afghanistan and their training under the Taliban and Al Qaeda further radicalized political Islam in the country.

Recurrent genocidal violence against the Hindus by communal sections of the Muslim community has forced hundreds of thousands of Bangladeshi Hindus to flee to India.

From 31 October to 2 November 1990, the Ershad administration engineered large-scale atrocities against the Hindu community in Chittagong and Dhaka cities in the wake of a fabricated press report that the Babri Masjid had been demolished. It was a stratagem to divert public attention from the all-party mass movement aimed at ousting Ershad from power.

In 1991, the Bangladesh Nationalist Party (BNP) came to power with the support of the fundamentalist Jamaat-e-Islami at the crest of a no-holds-barred anti-India campaign. In December 1992, when the Babri mosque was actually destroyed, the BNP and Jamaat cadres jointly perpetrated massive unrestrained Hindu cleansing operations (murder, rape, arson and looting) throughout rural Bangladesh.

With the BNP–Jamaat coalition coming to power in October 2001, the Hindu community throughout Bangladesh was once again subjected to an avalanche of murder, rape, looting arson and destruction and desecration of Hindu temples and deities.

The BJP-led National Democratic Alliance Government followed a soft policy towards the incumbent regime in Dhaka in the hope of tapping the latter's huge natural gas reserves and securing transit and trans-shipment facilities for the north-eastern states through Bangladesh territory. To that end, New Delhi maintained a studied silence over the genocidal atrocities on the Hindus while the Khaleda Zia Government looked the other way. As a result, persecution of religious and ethnic minorities (Hindus, Buddhists, Christians, Ahmediya Muslims, Chakmas, Garos, Hajongs and Santals) became a permanent and ongoing feature in the country, forcing tens of thousands of people to flee to India on a continuing basis.

Emergence of a whole range of Islamic terrorist groups linked to Al Qaeda like the Harkatul Jihad-e-Islam (HuJI), Jamaat-ul Mujahideen, Sahadat Al Hikma, Jagrata Muslim Janata and Hijbul Tauhid – and comprising mostly the alumni of over 6,000 Jamaat-run quomi mosques – have wreaked havoc in Bangladesh. In a series of bomb and grenade attacks these outfits have killed a number of secular opposition leaders and liberal intellectuals. The killing of a former Finance Minister in the Sheikh Hasina Government and an Awami League sitting MP, and four others at a public meeting at a village in Habiganj district on 27 January 2005, was one such incident in a long series of major bomb and grenade attacks in the last five and a half years that has taken toll of hundreds of lives. None of these cases of terrorist attacks have been resolved because, instead of properly investigating the crimes, the government used them as a tool to persecute Awami Leaguers and human rights activists, and to malign India. At a joint meeting in May 2002, representatives of these Islamic terrorist organizations of Bangladesh and the Minorities United Liberation Tigers of Assam (MULTA) and the Rohingiya Solidarity Organization (RSO) resolved to coordinate their activities in order to establish a Talibanized transnational Islamic state comprising Bangladesh, Assam, Tripura, the Muslim majority districts of West Bengal and the Rohingiya Muslim-dominated Araka Hills of Burma.

Illegal immigration

Unemployed agricultural workers, mostly Muslim from the economically lean rural districts of undivided Bengal, traditionally migrated to Assam in search of better living conditions before the partition. In the post-independence era, recurring communal riots in East Pakistan endangered the life and property of the religious and ethnic minorities and forced hundreds of thousands of families to cross the border into West Bengal, Assam, Tripura and, to a lesser extent, to the north-eastern states. The rise of radical Islam in the post-1975 era has heightened the insecurity of the minorities and the secular Muslims in Bangladesh. Discrimination against the minorities in matters of employment and dispossessing them of properties by force or fraud using the Vested Property Act (earlier called Enemy Property Act) in collusion with the administration, greatly increased the flow of migrants to India.

Table 12.1 Decline of Hindu population

Year	*% of Hindu population*
1941	28
1951	22
1961	18.6
1974*	13.5
1981	12.1
1991	10.5

* There was no census in 1971

The situation in Bangladesh is such that the country's minorities have the option either to migrate en masse to India or embrace Islam. Until now they have opted for migration, though conversion to Islam is also on the increase. This is evident from the sharply declining Hindu population during the period 1941–2001.

At this rate, in the next 30 years the Hindu population will cease to exist in Bangladesh, turning the country into a monolithic Islamic state. Since the outbreak of famine in Bangladesh in 1974, pauperized Muslims in ever increasing numbers have been clandestinely migrating to India. According to a rough estimate, currently the ratio of Muslim–Hindu migrants crossing over India and settling in this country is 3:1. So far not less than 15 to 20 million Bangladeshi nationals have migrated and settled in India. While the majority of them have settled in border states like Assam and West Bengal, others have moved on to other states and metropolitan cities like Delhi and Bombay. The details of the known state-wise distribution of Bangladeshi migrants are shown in Table 12.2.

The concentrated presence of Bangladeshi immigrants has radically altered the demographic structure and communal composition of the population of the border districts of the states adjacent to Bangladesh, particularly Assam and West Bengal. Major crimes like robbery, rape, murder and circulation of fake currency committed by criminal gangs from across the border with the connivance of the immigrants have registered a sharp increase. Smugglers, gun-runners, bottleggers and drug peddlers dominate the border

Table 12.2 Distribution of Bangladeshi migrants

States	*Numbers in millions*
West Bengal	5.4
Assam	4.0
Bihar	0.5
Delhi	1.5
Tripura	0.8
Rajasthan	0.5
Maharastra	0.5

belt. The rising sense of insecurity has been forcing many Hindu inhabitants in the border areas to move out to urban areas, disposing of their homesteads and landed property to the Muslim immigrants from Bangladesh for a song. Keeping pace with the changing contours of the communal composition of the population, mosques and *madrassas* have proliferated in the border districts, boosting Islamic fundamentalism. Lately, Muslim fanatics in Murshidabad district issued *fatwa*, subjecting members of the Baul community (Muslims) who preach universal humanism through their music to economic and social boycott with impunity.

Patronized by political parties, particularly in Assam and West Bengal, the immigrants have acquired all the trappings of Indian citizenship – ration cards, immovable property and the rights of franchise. Many have joined government services including defence establishments, police and paramilitary forces and security agencies. In Assam, by virtue of their numerical strength, they determine the election results in 56 of the total 126 state assembly constituencies. In West Bengal they also enjoy comparable political influence.

The demographic invasion has eroded the sovereignty of the Indian state, which has no say over who comes and stays in the country. It has effectively shifted the Indo-Bangladesh border inside India. The long-term danger of this phenomenon to the security and territorial integrity of India has to be viewed against the Greater Bangladesh demand raised by Maulana Bhashani in the late seventies, and the demand of *lebensraum* in the north-east region for the teeming millions in Bangladesh voiced by a section of Bangladeshi intellectuals in the early nineties. Deportation of 20 million immigrants is impossible, particularly because the Bangladesh Government refuses to countenance it, bluntly denying the very existence of the problem.

There is significant migration from Nepal to Assam, Meghalaya and West Bengal. Nearly 20,000 Nepalese have left Bhutan and taken shelter in India. Some 30,000 refugees from Burma, mostly Chins, have taken shelter in Mizoram. Such migrations increase competition for local employment, which generates racial tensions. The presence of the Chin refugees in Mizoram has triggered hostile actions against the migrants. About 25,000 Bangladeshi migrants staying in India along the India–Bhutan border commute daily to Bhutan for work.

Crisis in north-east India

The problem of cross-border terrorism and ethno-religious tensions in India's north-eastern states is a Pakistan era legacy. Not reconciled to the inclusion of Assam and Tripura in the Indian Union, Pakistan supported insurgencies by the Nagas and the Mizos since the mid 1950s and mid-1960s respectively. It stoked a low-intensity war against India by providing them with logistic support, training, arms and base facilities in the Chittagong Hill Tracts of the then East Pakistan.

The liberation of Bangladesh significantly reduced the stream of ethnic insurgency for some time. But the rise of the ULFA in the mid-1970s gave a renewed thrust to ethnic insurgency in the region, with the Kachin rebels in Burma providing safe havens and base and training facilities to the insurgent outfits in areas under their control. In 1990, when the Kachin Independence Organization (KIO), acknowledging New Delhi's strong support to the democratic movement in Burma, sent the ULFA, the National Socialist Council of Nagalim (NSCN) and PLA contingents back to India, the Ershad Government gave them shelter and training facilities in Dhaka city and the Sylhet and Maulavi Bazaar districts. The ISI, in coordination with the Bangladesh Directorate General of Forces Intelligence (DGFI), arranged guerrilla training for successive batches of ULFA cadres in Pakistan. Dhaka provided the necessary staging facilities and travel documents. Later, other insurgent groups like the National Democratic Front of Bodoland (NDFB), All Tripura Tribal Force (ATTF) and National Liberation Front of Tripura (NLFT) availed themselves of the hospitality of the Bangladesh Government.

The first Khaleda Zia regime (1991–96) followed Ershad's policy of supporting the north-east insurgents with Begum Zia, eulogizing them as freedom fighters. The ULFA and NSCN (I/M) imported several consignments of arms on behalf of all the north-east insurgent outfits from South East Asia by sea through Cox's Bazaar coastal belt and thence transported them across country to their operational bases in India through the Chittagong Hill Tracts with the logistic support given by the DGFI. Paresh Baruah, the ULFA army chief and a number of other ULFA leaders have been living in Dhaka since the early 1990s. Anup Chetia, the convicted ULFA general secretary, guides and organizes terrorist operations in Assam from the Dhaka central jail through remote control. The ATTF and NLFT have been operating inside Tripura from their bases in the Brahmanbaria and Habiganj districts in Bangladesh, frequently killing or kidnapping Bengalis with the support of the DGFI. Requests from the Government of India to dismantle the insurgent camps have met with the stock cynical reply that there is no Indian insurgent presence in Bangladesh.

In the 1990s for the sake of operational convenience, the ULFA established new bases in southern Bhutan and used them as a launching pad for cross-border terrorist attacks in the lower Assam districts. In Bhutan, the ULFA trained NDFB and Kamtapur Liberation Army (KLA) cadres in guerrilla warfare and provided them with operational logistics. It motivated sections of the Nepalese refugees from Bhutan into setting up the Communist Party (Maoist) of Bhutan on the lines of the Maoist rebels in Nepal. ULFA has been involved in importing arms through Bangladesh and sharing them with the Maoist insurgents of Nepal.

Although the Government of India's ceasefire agreement with the NSCN (I/M) has weakened the ethnic insurgency in the north-eastern states, the Naga rebel group's demand for incorporation of the Naga-inhabited areas of

Manipur and Arunachal Pradesh in a greater Nagaland remains a sticking-point in the peace negotiations. The Bhutanese army crackdown on the ULFA bases in the Himalayan kingdom in December 1993 hit the insurgent outfit hard in terms of loss of combatants and weapons. This, and the fact that ULFA leaders have made massive investments in Bangladesh, have increased the dependence of the ULFA on the support of Bangladesh and Pakistan. On receiving fresh supplies of arms with the support of the DGFI, the ULFA has been making renewed strikes in Assam and Nagaland. Subsequent initiatives for peace talks with the government floundered in the face of the ULFA's insistence, following the direction of the DGFI, for inclusion of Assam's sovereignty in the agenda for discussions.

North-east insurgents' links to Burma

Burma came to be used in the late 1950s as a route for the Naga insurgents to travel to Yunnan province in China for guerrilla training from the Chinese. In their long and arduous treks to China, the KIO provided them the much needed staging facilities. Later Mizo rebels also travelled to China for training and arms through Burma. In the post-Mao era, when the Chinese Communist Party stopped exporting revolution, the NSCN rebels fell back to the KIA Second Brigade areas for shelter and training. The ULFA and Meiti insurgents also joined them there in due course. In 1988, following a bloody clash between its two rivals, the Naga groups in the Kachin state, the NSCN split into NSCN (Issak/Muivah) and NSCN (Khaplang). The continuing tensions between the two NSCN factions have impeded the progress of peace negotiations in Nagaland. In 1990, when the KIO asked the north-east insurgents to leave Kachinland, most of them trekked to Bangladesh, but small contingents of ULFA, Meiti and Chin rebels (Burmese) stayed on at the NSCN (Khaplang) headquarters on the western bank of the Chindwin river in the Patkai range of Burma.

As a part of the strategic cooperation between Burma and India, the Burmese army with its Indian counterpart, carried out joint counter-insurgency operations with marginal success. During the state visit of General Than Shwe, the Chairman of the State Peace and Development Council to India in October 2004, the two countries agreed on closer strategic co-operation, but there were two limitations on the Burmese Army's ability to adhere to this. First, inadequacy of combat force and weapons deployed in the Sagaing division to jettison the trans-border movements of the insurgents and secondly, corruption in the Burmese army is rampant. In the 1995 jointly conducted operation 'Golden Bird', the armies of the two countries encircled a mixed group of Indian insurgents carrying arms cross-country, killed thirty-eight rebels, captured more than 100 pieces of arms and arrested ULFA foreign secretary Sasha Chaudhury, but after 12 days they mysteriously allowed the remaining 100 rebels to escape with considerable weaponry. Again, in November 2000, the Burmese army captured 192 Meiti guerrillas of Manipur, but did not turn them over to the Indian authorities. After holding them for three months, the Burmese released them in February 2001 when the rebels paid the Western Command of the Burmese army a bribe of three crore rupees. Gen. Soe Win, Than Shwe's new Prime Minister, was then chief of that command.

Continued military rule in Burma has isolated the country from the rest of the world. The ruling junta smashed the pro-democracy uprising in Burma in 1988, killing thousands of demonstrators and incarcerating many pro-democracy political activists including Aung San Su Kyi and other National League for Democracy (NLD) activists. The US, Western Europe, Japan and other major countries imposed economic sanctions on Yangon (Rangoon) and stopped giving aid. China, however, continued to benefit from its growing trade with the repressive military regime and supplied military hardware. India fully supported the democratic movement and distanced itself from the military government. It strongly denounced the brutal crushing of the uprising, and offered refuge to political exiles.

Under international pressure, the junta held a general election in 1990. Though the National League for Democracy (NLD), led by Aung San Su Kyi, won a landslide victory, the military regime has refused to hand over power to the elected representatives of the people. Against this backdrop, beginning in 1993, on the pragmatic consideration of trade and security interests and to contain the growing influence of China in Burma, India has reversed its policy of support for the movement for democracy in the country and established close ties with the junta.

As noted earlier, joint counterinsurgency operations have yielded very little concrete result. In the last few years, India has made significant investments in the road, railway and communications sectors in Burma and given trade concessions to that country, but New Delhi has a 1:3 trade imbalance vis-à-vis Yangon. The junta has concentrated on improving its control over its borders with China and Thailand and paid little attention to its frontier with India. Close Burma–China military co-operation provides weapons and other combat equipment.

Ethno-religious tension

The Indian north-east region, comprising a territory of 300,000 km^2 with a population of 46 million spread over eight states, has over five hundred different ethnic groups (110 of which are major) and consequent ethnic sub-national strains focused on traditional and primordial values of 'Tribe-land-Autonomy'. Massive immigration from outside the region has created an acute identity crisis among the native tribal populations accentuating ethnic militancy and at times triggering religious violence in India's troubled north-east peripheries.

The massacres of Bengali Hindu migrants at Mandai in Tripura in June 1980 by tribesmen, and of Muslim migrants at Nellie in February 2003 by

Assamese Hindus, were symptomatic of the undercurrent of ethno-religious tensions inherent in the sociopolitical and socioeconomic dynamics of the region. Land passing into the hands of the immigrants has been the main factor stoking the 'local–outsider', conflict that has found expression in agitations from time to time against the Bengalis and the Nepalese in Meghalaya, against the Bengalis and, more recently, the Biharis in Assam, against the Chins in Mizoram and the Chakmas in Arunachal Pradesh. In Manipur, intertribal conflicts, particularly bloody Naga–Kuki clashes, are rampant and have taken a massive toll in the state.

Supplementing the efforts of Christian missionaries, the NSCN played a proselytizing role among animist Naga tribes of Arunachal Pradesh and western Burma, raising the slogan 'Nagaland for Christ' as the means to expand the area of the future sovereign Naga state. In Arunachal Pradesh, there were only 1,710 Christians in 1961, but their number has increased manifestly to over 125,000 now. Presently there are 120,000 Christians in Tripura, a 90 per cent increase since 1991. There is a direct link between conversion and secession. Tribal Hindus account for about 22 per cent of Tripura's 3.2 million people. Outlawed NLFT militants have been forcibly converting tribals to Christianity. In the remote areas of Tripura, vigilante groups have been formed to resist forced conversion.

Water sharing dispute

It is in the twenty-first century that wars would be fought over sharing of water resources. Growing inadequacy of fresh water would most certainly define the future conflicts in South Asia. Construction of the Farakka barrage by India triggered the water-sharing dispute with Bangladesh and a sustained anti-India campaign in the country, accusing India of desertifying large areas of the country. Despite the signing of an agreement in 1996 for sharing Ganga water, the issue remains potentially volatile, although under the said agreement Bangladesh gets 100,000 cusec more water than its entitlement. In the lean season, at the expense of the Calcutta port, the BNP–Jamaat Government wants to raise the issue at international forums and has launched a campaign against India's plan to route water from water-scarce to water-stress regions through an ambitious river linking project.

The Indian stand on management and sharing of water resources is based on the theory of 'natural sovereignty', whereas Bangladesh believes in the theory of 'natural flows'. On the issue of sharing the common rivers of India, Bangladesh and Nepal, and the related question of augmenting their flows, Bangladesh wants trilateral talks as a ploy to stall Indian initiatives in the matter. Pakistan's objection to India's Baglihar Hydro Electric (450 MW) Power Project on the Chenab river and to the construction of the Kishanganga power project on the Jhelum would inspire Dhaka to intensify its anti-India campaign on the river-linking issue.

Dhaka's unrelenting intransigence over critical bilateral issues like illegal immigration and state-sponsored cross-border terrorism in the north-east will continue to bedevil Indo-Bangladesh relations. Short of a miracle, the chances of an early breakthrough are remote. Conflicting ethnicity and religion, land and language, population movements and insurgencies define the multi-faceted security crisis in the region. Proliferation of arms and drugs has heightened the crisis, undermining the region's economy and sociopolitical stability, and straining its bonds with the rest of the country.

Thriving arms bazaar

The Vietnam war and the post-1975 conflict in Indo-China fed the arms black market in Thailand and Burma. In the sixties and through the late seventies China supplied a lot of weapons to the Naga and Mizo rebels. Pakistan supplied them US-made weapons at their bases in the Chittagong Hill Tracts. According to an estimate of the Indian military intelligence, the Naga insurgents received not less than 3,000 pieces of assault rifles, automatic carbines, light machine guns and rocket launchers. Approximately 80 per cent of the weapons were recovered or seized, lost in action in India and Burma, and deposited by the surrendering rebels since the 1960s. Some of the weapons of the Mizo National Front found their way into the black markets in Bangladesh and north-east India. The Bangladesh liberation war in 1971 caused a major spurt in small arms proliferation in that country and north-east India. Many members of the liberation force did not surrender arms after the war despite Sheikh Mujib's appeal to them to do so. Those weapons were later sold in the underworld arms bazaar. Elements in the Indian army also sold off many weapons seized from the defeated Pak army.

In the last decade of the twentieth century, the north-east rebel groups procured from Thailand roughly around 10,000 assault rifles, carbines, pistols and revolvers, grenade-firing rifles and an assortment of other weapons. The weapons were shipped from Thailand to Cox Bazaar coastal belt in Bangladesh and, as noted above, thence shifted cross-country to the rebel operational bases in India. After the Burmese army disrupted this route, the insurgent outfits have been collecting arms from the Blackhouse and Ah Hua networks of Yunnan. For the last several years over 3,000 pieces of small arms were collected from these sources, but more than 1,600 of these weapons were seized by the Burmese troops during raids on Manipuri rebel bases around Tamu in November 2001. Seizure of arms has been frequently reported from the Bandarban district at the India–Burma–Bangladesh tri-junction and Chittagong. The seizure of ten truck-loads of sophisticated weapons from the Chittagong port area in 2004 and a truck-load of ammunition near Bogra city pointed to weapons being imported from abroad both by road and by sea and then sent to the operational bases of north-east insurgents by road.

The drug trail

Notwithstanding the continued production of drugs in the Golden Crescent area of Afghanistan, the importance of the Golden Triangle in South East Asia (comprising parts of eastern Burma, northern Thailand and western Laos) as the main source of production of and trading in drugs remains. As a result, the Indo-Burma border has emerged as major heroin trafficking area and heroin has now become Burma's most valuable export. Since 1988 opium production in Burma has risen to over 2,030 metric tonnes annually, amounting to 60 per cent of the total world supply. Heroin from Burma meets the demand in North America and Australia. Over the past few years Burmese heroin trafficked out of north-west Burma through north-east India has been serving the European heroin market although heroin sources in Europe originated from the Golden Crescent, Pakistan, Afghanistan and Turkey.

Traditional druglords like Khun Sa have been eclipsed by ethnic rebel armies including the United Wa State Army in the Triangle. After the once strong Burmese Communist Party (BCP) withered away in 1998, its Wa officers took to drugs. Lately, the UWSA has monopolized the methamphetamine production so much that a *Time* magazine cover article described the Was as the 'Speed Tribe'. To retain their control over production and export of heroin to Laos and Thailand, drug lords like Khun Sa have imposed a 60 per cent 'profit tax' on smaller cartels, forcing at least three of them – headed by Zhang Zhi Ming (former Burmese Communist Party official), Lo-Hsin Nian and the Wei brothers – to relocate their 14 to 18 drug refineries to the borders with India's north-east, mostly in the Sagaing division, Chin Hills and the Arakans.

These cartels use nearly 30 routes to push their ware into north-east India on the way to the Western markets which have been identified by the Narcotics Control Bureau (NCB). While the routes through Manipur and Mizoram have been used for two decades or more, the ones through Nagaland, Arunachal Pradesh and Tripura have just come into use. Frequent seizure of heroin has been reported from the north-eastern states. In January 2002 alone, police and customs officials seized nearly 3 kilograms of Burmese heroin and more than 10,000 methamphetamine tablets in Mizoram, and 1.5 kilograms of heroin in Assam and Tripura. The arrested traffickers confessed that drugs were on their way to Bangladesh, where lax anti-narcotics laws have encouraged Indian and Burmese drug dealers to use that country to ship their deadly cargo to the West and the Far East. This spurt in drug trafficking through the north-east poses a serious threat to the region and the country.

Trafficking through the north-east has sharply increased the population of drug addicts that is now estimated by the Indian Council of Medical Research at 110,000. The number of HIV positive cases in the north-east has risen to 12,000 over the last two decades. Manipur and Mizoram have been the worst affected – more than 1,650 people, mostly youths, have died of drug-related maladies. This drug addiction could slowly spread to all the states of the north-east and affect the youth and social fabric of the region. The involvement of serving military and paramilitary personnel in drug trafficking is on the rise, creating problems of discipline and morale in the Indian armed forces and weakening the policing of frontiers and the drug trade.

Ethnic rebel armies in the north-east who, in the past, had resisted production and trafficking of drugs, are beginning to display the 'Burma syndrome'. While some of the rebel groups like the Manipur Peoples' Liberation Front continue to resist the drug traffickers, meting out exemplary punishment to them, other groups are turning to taxing drug mafias to raise funds. They are also encouraging tribal farmers to plant poppies, acting as agents of the Burmese drug lords who have relocated themselves in western Burma.

Despite clear indications that the Burmese drug mafia are increasingly using the north-eastern states of Bangladesh, mainland India and Nepal on their way to the global market, Indian narcotic control officials have tended to play down the threat. India has lately adopted a 'Look East' foreign policy, but has not joined the Special Task Unit formed in 2002 by China, Thailand, Burma, Laos, Cambodia and Vietnam. The emerging rebel-drug mafia-officialdom nexus could undermine the official machinery in the sensitive north-east frontier region, which India can ignore at its own peril.

Conclusion

Effective border management and surveillance geared to eliminating illegal cross-border movements of goods and people is the key to homeland security and has to be at the core of all programmes for deterrence. The existing arrangement for border surveillance through border patrolling, deployment of observation sources along the borders and aerial surveillance by the ARC have not been effective, primarily because of the incompetence and indifference of the agencies concerned in putting the existing resources to maximum use.

Western countries use a comprehensive network of surveillance radars, environmental sensors and communication systems to keep a tab on the activities at their respective frontiers. In the context of infiltration into Kashmir, the government has embarked on a composite programme of border fencing and placing multimode sensors. For the eastern sector, the crying need is to implement the border-fencing project on the Bangladesh frontier. River and coastal patrolling also needs to be augmented. Once these basics are taken care of, introduction of multisensor surveillance systems can be considered for the eastern sector in phases based on the experience in the western sector.

India needs to take effective steps to stop further illegal immigration by plugging the loopholes in border management, like corruption in the Border Security Force, by removing, as far as practicable, the incentives like the issuing of ration cards, deletion of migrants from the voters list, curtailing employment opportunities, prosecution under the Foreigners Act, denial of social security facilities, etc.

An exhaustive survey and identification of illegal immigrants and their locations needs to be undertaken on an urgent basis. The refugees among them who have migrated in the face of religious persecution should be naturalized case by case and the rest should be declared non-citizen residents. Such of those non-citizens who cannot be deported should be dispersed from the border areas to less sensitive regions and permitted to work.

Immigrants come mostly from economically lean rural areas. International organizations and donor countries have to be persuaded to earmark an appropriate percentage of their aid to Bangladesh for funding employment and generating projects in the immigration-prone areas of the country.

The Government of India should use all its diplomatic prowess to bring effective international pressure on the Bangladesh Government to ensure the safety and security of the religious and ethnic minorities to live in peace and dignity in their homeland, and has to use all its political, economic, diplomatic and strategic leverages vis-à-vis Bangladesh to that end proactively. The international community has to be convinced of the need for humanitarian intervention to protect the minorities of Bangladesh as was done in Bosnia, Kosovo and Somalia, if other measures fail to achieve the objectives.

13 Security of Himalayan frontiers

Role of science and technology, modern air surveillance, remote sensing

Vinod Patney

Introduction

The mighty Himalayas stand as a sentinel to dissuade would-be aggressors into India. The difficult terrain, rarified atmosphere, and extreme cold present unforgiving challenges. Our experience in Siachen has shown that the environment is probably more deadly than the enemy. The psychological effects and other illnesses degrade soldiers' ability. The accuracy and performance of weapons also suffer. For instance, at an elevation of only 3,100 metres, a round fired at a range of 1,000 m will impact as much as 180 cm higher than at sea level.

'High altitude' has been defined as being between 2,500 and 4,800 m and the term 'very high altitude' is used for heights above 4,800 m. Beyond 4,900 m, human beings can function only for short periods of time. The Himalayan range contains areas of high or very high altitude. It is the highest mountain region in the world with more than 30 peaks above 7,700 m. It extends over an arc of 2,410 km from the river Indus in the west, and runs across Kashmir and northern India, the southern part of Tibet, most of Nepal, Sikkim and Bhutan, and on to the eastern regions of India. It is indeed a formidable obstacle.

Yet, for the invaders, 'Fortress Himalayas', though a major challenge at all times, has not been impregnable. History records a number of successful crossings from the Aryan invasion of 1500 BC and Alexander's march into India in the fourth century BC, to the recent major infiltration that led to the Kargil conflict in 1999. The looming presence of the Himalayas should not make us complacent.

The present and expected geo-political situation suggests that we must remain alert. The inherent differences between India and Pakistan are unlikely to disappear in the near future; the general view is that the differences are endemic and that it is likely that Kashmir will remain a stumbling block. As for China, the border dispute has yet to be settled and it has been argued that a conflict of interests could take on serious proportions in the future. China has launched a massive programme to develop road and rail communications and telecom infrastructure to facilitate progress in the integration of Tibet and peripheral territories. We should view these developments as significant.

The products of modern technology have considerably overcome the problems of high altitude and unforgiving terrain. The rapid progress in different aspects of air power has ensured that the Himalayas are no longer a barrier. The actual business of fighting war has also become more efficient and effective. All this has resulted in a sea-change in the nature of conflict and conflict resolution.

Although the primary purposes of military forces to deter and to coerce have remained, diplomacy and economic issues are now playing an increasing role in effecting both deterrence and coercion. The nature of conflict is now well beyond military confrontations only. With globalization, increased competition and the shrinking of the globe, conflicts are no longer 'events' with a clear start and end. In fact, we are now in the realm of continuous conflict interspersed with the sporadic clash of military forces. Conflict resolution has also become multidimensional, and with increasing interdependence of nation states, the chances of a major military conflict have receded unless considerable asymmetries in capabilities exist. Thus, whilst adequate military capability will continue to be the final arbiter that underpins national security, the salience of the armed forces in conflict and conflict resolution is likely to decrease provided we have the capability to deter aggression effectively.

In the strictly military domain, information and its analysis in time is the premier requirement. The relevance of the phrase 'information is power' is increasing in all fields of activity. It should be taken for granted that technological capability exists to hit, damage and destroy any target anywhere in the world as long as it can be properly identified. The crucial issue is timely intelligence on the adversary and the means to analyse and use the information. Pre-warning and effective intelligence are even more important in the vast Himalayan region, a region that is near impossible to guard by the physical presence of troops alone. This paper seeks to evaluate the role of science and technology and remote sensing techniques in aid of intelligence gathering, and utilization of such intelligence to enhance security.

Gathering of intelligence

Intelligence gathering is a continuous activity, as important during times of 'relative peace' as during conflicts. The requirements range from information to judge the possible intentions of the adversary, to inputs that could help to seize opportunities for coercion or to further our interests. During actual military hostilities, including the lead up to hostilities, the need for accurate and actionable intelligence becomes more urgent. In mountainous terrain, it takes time to concentrate forces for offence or defence, and intelligence gathering must be time-sensitive and cover an extended area.

The epitome of intelligence gathering for military operations is that every object or electromagnetic radiation of military relevance is accurately located and characterized, as often as necessary, to give an indication of the threat, or an opportunity for an offensive that the situation may afford. Remote sensing is used to garner the required information.

Remote sensing has been defined as 'the science (and to some extent art) of acquiring information about the earth's surface without actually being in contact with it. This is done by sensing and recording reflected or emitted energy and processing, analysing and applying that information.' To this definition we could add 'the acquisition of information about airborne objects and objects in space'. The sensors could be active or passive, ground based, airborne or in space. The types of ground based sensors include air surveillance radars, battlefield surveillance radars, acoustic sensors, infrared sensors, low light intensifiers, etc. Ground based sensors have limited range, and as a large number would be required for even small areas, command and control aspects become somewhat complicated. Airborne and space based sensors are more versatile.

Remote sensing is required to provide:

- Area surveillance for ground and airborne objects. Very soon, location and functioning of enemy satellites will become required information.
- The necessary surveillance could be periodic or at times continuous. The information could also be required by day and night and in all weather conditions. Most remote sensing is based on line of sight, and mountainous terrain will cause masking and shadow areas. Shadow areas decrease as the sensors are placed at greater heights and as they approach the near vertical over the target.
- Detection, location and classification of enemy emitters that would include radars and voice or data communications.
- Target detection for targets on the ground or airborne.
- Target location and motion, again on the ground or in the air.
- Target classification and recognition/identification as 'friend or foe'.
- Post-strike damage assessment that could include a degree of collateral damage, if any.
- Data needed for production and updating of maps and other military applications like simulation, training, weapon guidance, etc. Weapons whose accuracy is based on terrain matching radars or terrain comparison are dependent on information acquired by remote sensing. Again, in the lead up to the 1991 Gulf War, the maps used by the allied forces were out of date and inaccurate. Better and up-to-date maps were a requirement given the featureless desert terrain. In just a few months, using satellite imagery, 4,500 different maps were produced. Before 'Desert Shield' became 'Desert Storm' as many as 35 million updated maps were distributed to the allied forces.

For sensing electromagnetic emissions and airborne objects, appropriate radars and sensors are required. The capability of such sensors is dependent on power, range, sensitivity and line of sight requirements. There are no night or weather restrictions as long as the sensors are in position. Admittedly, there are measures that could be applied, but then there are countermeasures as well. This adds substance to the adage that for every offensive system produced, a defensive system will be put in place and vice versa. Radar and missile warning receivers and laser warning systems are part of the 'defensive' avionics suite. Offensive Electronic Counter Measures (ECM) use broadband or discrete jammers as well as decoy systems. Electronic Support Measures (ESM) is the term used to monitor radar and communication signals, and analyse such signals to determine geo-location and source identification. ESM equipment is also needed to follow changes in the frequency spectrum adopted to retain security of information.

Large strides have been made in making the sensors far more sensitive, and indeed in the entire gamut of information warfare. However, it is unlikely that foolproof systems will be fielded in the foreseeable future. Stealth technology is also developing rapidly and the concept of stealth is not limited to detection by radar, but includes low acoustics and low observables as well.

Remote sensing of ground terrain can be done photographically or electronically. Photography has limitations as the photo plates or rolls have to be developed on the ground before they are analysed. This can take considerable time. In fact, during the Kargil operations, the time taken for the reconnaissance mission to be launched, the mission itself, post-flight developing, processing, analysis and dissemination was far too great and, even then, the results were not of good enough resolution to be able to discern the type and/or location of targets. An innovative method was adopted where a pilot carried a normal video camera and filmed the target area. The digital film could then be analysed immediately on landing, with magnification to give much better clarity and resolution. Target detection, though still difficult given the size of targets, was made more efficient.

Digital recording systems are definitely preferable as, apart from other advantages, the results are easier to process. However, photography of any type will always be dependent on light conditions and transparency of the intervening atmosphere. Other methods of photography involve use of low light amplification and infrared cameras. Infrared cameras can be used by day or night, but as they are dependent on differences in temperature, the resolution is lower particularly during and after precipitation or through cloud cover. Active systems that use a laser source to illuminate the target are under development, but laser systems are adversely affected by hazy conditions.

Infrared cameras do have a limited capability to see through camouflage, but the new technology of hyperspectral imagery is a major development. Reliable classification and identification of non-radiating targets is a challenge and so is the problem of differentiating between a target and a decoy. Hyperspectral imagery will help detect targets of low radar cross-section, or which are wholly or partially hidden by foliage or other camouflage, or those that have a low contrast as compared to the background.

Hyperspectral imagery is imaging stereoscopy where hundreds of low bandwidths are used to record images. This gives a spectral response that carries far more information than a mere picture. The spectral signatures are different for each element and, therefore, the system can see through camouflage, concealment and decoys. The camera senses the nature of the material and not its physical form. Hyperspectral imagery does need some target illumination from a light source but it is less affected by shadows. The system may not be able to picture soldiers on the ground but it can locate areas of disturbed soil, such as freshly laid minefields or other earthworks. The infrared portions of the spectrum could also be used but not to a great extent as target differentiation and imaging will require multi-spectral bands that are most appropriate for the task.

The one sensing system that is unaffected by weather or time of day is Synthetic Aperture Radar (SAR). SAR can produce images akin to photographs but cannot see through camouflage and decoys. However it does give fairly accurate definition of form and outline and can pick up metallic objects like fencings, etc., very clearly.

It will be seen that there is no particular system that represents a panacea, but each has some advantages over the others. A combination of systems is needed. A similar situation obtains with respect to sensor platforms.

Sensor platforms

The different types of airborne sensor platforms include static platforms like balloons and aerostats, aircraft that operate at varying speeds and heights, Unmanned Aerial Vehicles (UAV), and satellites. It is unlikely that any one type of platform will carry all types of sensors.

Aerostats

Aerostats or tethered balloons have the singular advantage of being able to stay on station for considerable periods of time and hence effect surveillance of a designated area continuously as long as they are held aloft. However, they are affected by strong winds and have to be lowered for maintenance at regular intervals. As they are static, their location cannot be kept secret and they are vulnerable to enemy fire from the air or from the ground. For this reason, they should not be positioned too close to the border and thus the effective look into enemy country is reduced. The height to which the aerostats can be raised is limited and finite, and this fact also reduces the area under surveillance. Most importantly, in mountainous terrain, even if it were possible to locate an aerostat safely on a hilltop, the impact of terrain masking will be too severe for it to be a worthwhile sensor platform.

Manned aircraft

Helicopters can also carry sensors but, once again, they are vulnerable and the height restrictions lower their possible utility. Faster aircraft are less vulnerable but they are only good for a single pass type of reconnaissance. The longitudinal coverage is reasonable but the lateral coverage is a function of height. At lower heights, the coverage (swath area) extends to only a few hundred yards or less. However, the image resolution is good and such low level passes are indicated when the target location is well known and a more detailed picture of a small target is required. In mountainous terrain, such targets will be few and far between. The difficulties of accurately pin-pointing the target should not be underestimated.

At higher heights, the area coverage increases – at the expense of resolution – but the problem of masking and shadows still persists. Some good information can be acquired but the process is not troublefree. The limitations of terrain masking are reduced if the aircraft were to fly directly over the target. Effective surveillance over the area of concern would imply flying over the territory of the adversary, which may not be permissible when a no-war situation obtains. In any case, high-level missions can seldom be kept secret. Another disadvantage of aircraft platforms is that they cannot stay over the target area for too long, whereas repeated missions are necessary to develop a picture of possible enemy intentions.

Unmanned Aerial Vehicles (UAVs)

UAVs are versatile machines. They can be made light and small and can stay on target for extended periods of time. UAVs fly at low speeds and can be vulnerable to enemy aircraft, but adopting low observable technology and reducing their acoustic signature can make them less noticeable. They are the preferred option when relatively long duration surveillance is required or over well-defended areas where manned aircraft would be at considerable risk. UAVs generally have interchangeable sensors and can operate by day or night. Information acquired by UAVs is transmitted via data links to wherever desired on real time basis. Hence, the external UAV pilot can manoeuvre the UAV for a second or third look over a particular area, and cause it to descend if necessary for better resolution. In this manner, interactive surveillance is possible. This is a great advantage. The range of UAVs is dependent on line of sight distance, but with relay stations, the range can be extended almost indefinitely. During the recent war in Iraq, Global Hawk UAVs were 'piloted' from mainland US.

Previously UAVs were perforce restricted to lower altitudes and consequently could be flown out to limited distances only. With new technology now available, they can operate at great heights. The Global Hawk UAV has reached a height of 20,000 m and can operate for 30 hours at a stretch.

Research is continuing for UAVs to fly at even greater heights and for much longer periods.

With solar power, the UAV could fly for weeks or months at a time. One such UAV, the Helios, reached a height of 30,000 m on 13 August 2001. The possibility of using nuclear power for UAVs is under examination and if the technology is successful, they could stay aloft for many months and operate at still greater heights. At a typical altitude of around 28,000 m, the UAV would be out of range of manned aircraft and, from such heights, a very large part of the earth would be under constant surveillance. The limiting factor would be the availability of sensors of the desired sensitivity.

At the other end of the UAV capability are mini-UAVs that could have a wingspan of as little as 15 cm ('Black Widow') and weigh only a few ounces. The more usable ones have a wingspan of slightly more than 1 m and weigh about 2 kg. These are electrically powered, inexpensive and, therefore, expendable. One such UAV, the 'Dragon Eye', was used extensively over Iraq almost at individual soldier level.

Space satellites

The great advantage of satellites is that once operational, they continue to provide data as programmed for many years. As they operate from great heights, the area coverage is much greater. Again, with technological advances, more sophisticated sensing devices are available, and these permit good resolution even from such high altitudes. Resolutions of one metre or so are now considered as normal.

Low earth orbit satellites operate at heights of around 700 to 900 km and have a revisit time of approximately five days. Reconnaissance satellites follow sun synchronous orbits implying that every pass over an area is at the same time of the day, thereby obviating problems associated with shadows, etc. If more frequent revisits are required, the satellite can be manoeuvred for the purpose but additional fuel is consumed and the life of the satellite is reduced. Our Technology Experimental Satellite (TES), in current operation, can be manoeuvred by increasing or decreasing the height of the satellite. A 1 km increase in height moves the satellite 8 degrees westwards, and similarly a reduction in height will move the satellite eastwards correspondingly. Such manoeuvring should be viewed as a one-off solution; for revisits on a more regular basis, more satellites are required.

The state of the art technology permits 'imaging' only from satellites that are in low earth orbit. In fact, it will be a few years before SAR on board satellites are operational. However, low or medium earth orbit satellites can be used for navigation purposes; the Global Positioning System (GPS) constellation satellites operate at medium earth orbit. As for communications and monitoring of electromagnetic transmissions, all satellites, including those in geostationary orbit, can be and are used.

The perfect sensor platform

It will be seen that none of the sensor platforms represents a panacea. Where time is not critical and a watchful eye needs to be maintained over enemy territory, satellites should be the platform of choice. For longer duration surveillance over a designated area, the UAV comes into its own. For specific reconnaissance of a target, manned aircraft are preferable. Aerostats provide a continuous surveillance of a limited area around the aerostat. Even in mountainous areas like the Himalayas, satellites, manned aircraft and UAVs have a role to play, but it is only the satellites that can overfly enemy territory without it being termed as air space violation.

Significance of modern technology

It is an old axiom that the side which is able to better absorb the products of modern technology, will be the victor. The adage takes an added importance in present times because of the rapidity at which newer and better products are being introduced. Some important areas of technological progress are discussed in the paragraphs that follow.

Weapon systems

TV pictures of the war in Iraq showed the accuracy and effectiveness of the attacks. It was shown that a bomb could be lobbed through a particular window of a particular building in the centre of town. The Israelis have time and again hit an individual vehicle in a convoy of vehicles that were reasonably close to one another. In the Kargil conflict, we also carried out effective pinpoint bombing that changed the course of the war. Longer range stand-off weapons are available. Yet professional magazines are awash with information on newer and even more efficient weapon systems. Each of these weapon systems could be war-winning. In the case of mountainous terrain, some modifications will be needed, and as our battlefields are likely to be the highest in the world, much of the work will have to be done by indigenous research and development.

Space technology

The 1991 Gulf war has been christened as the first 'space war'. Since then the importance of space assets has increased considerably. Space assets are used for communication, reconnaissance, surveillance and navigation. The application of GPS has transformed the accepted levels of accuracy. Space assets are also being used for missile warning, and soon space based radars will carry out effective surveillance of any object on the ground, in the air, or in space. The importance of space assets extends to warfare and intelligence gathering over mountainous terrain as well. In fact, so great is the contribution of space assets that it is on the cards that space warfare will soon be a reality.

Miniaturization

The marvels of miniaturization are part of our everyday lives. Size is reducing but capability and capacity of equipment is increasing. The advent of nanotechnology will again transform our lives and the means and manner of warfare as well. For instance, if the sensors are made more capable and smaller, more sensors of different types could be carried on the same platform thereby effectively introducing force multipliers. Alternatively, the platforms can be made smaller and less expensive. As for satellites, micro-and nanosatellites weighing a mere 10 kg or so will soon be available. The small size would increase survivability and reduce costs. The cost of launching low-weight satellites into orbit will also decrease. It will then become possible to have sufficient satellites and launchers available on ground that could be made operational readily and quickly. All this will add up to the creation of effective space assets with almost inbuilt redundancy.

Data processing

The different types of sensors and sensor platforms complement each other. However, they produce results that are in disparate forms in terms of sources, areas, type of information, etc. A superimposition of information is necessary to give a more complete picture. The considerable amount of data that could be generated, particularly if surveillance is to be undertaken fairly frequently if not continuously, is far too great for manual interpretation or analysis techniques. Human expertise will always be needed but software that, inter alia, permits automatic target recognition and change detection will be very useful. With such software, human photo interpreters will be able to zoom quickly on to areas of interest and provide timely intelligence. Also, as computers would do the initial processing, the chances of missing out on significant information are reduced. The process involves human ingenuity and can never replace *humint* but data processing techniques will definitely support the gathering and analysis of intelligence information. In similar vein, the enemy electronic order of battle can be monitored on a near continuous basis, and the information will always be useful in determining possible enemy intentions.

Two other areas under the broad heading of data processing merit discussion. The first relates to the major work being done towards Network Centric Warfare (NCW). NCW is intended to enhance the situational awareness of all agencies and individuals involved in the use of arms. In essence, the epitome of NCW is the collection, collation, processing, analysis and dissemination of desired information wherever and whenever required and in the manner in which it is required. The system would put the entire military hierarchy on to a single grid. A two-way interaction between say a sensor platform and the user(s) would also be possible. As a result, the planning and conduct of operations would become far more efficient and effective. The decision cycle would be considerably reduced, as the requisite information on which decisions are to be based would be readily available. The uncertainty element will also be reduced. For instance, in the 2003 Iraq war, the allied ground forces were always able to elicit information on the location of enemy formations. Again, a two-way audio-visual interaction was possible between a pilot and troops on the ground to ensure that the right target was acquired. It is true that there is some way to go before NCW becomes fully integrated and part of military infrastructure but, undoubtedly, any advance along this particular road cannot but be beneficial. Interestingly, it has been mentioned in a report that the bill to introduce the hardware required for NCW to cover worldwide US responsibilities could be as high as 200 billion US dollars. The figure may appear to be much too high but the importance of the capability can be gauged from another reported story that the US is spending 11.3 billion US dollars annually on command and control aspects alone.

Military use of the electromagnetic spectrum is bound to increase, which will improve efficiency, but the process will also bring about another area of vulnerability. Adversaries will try and undermine the effectiveness of the other's data processing systems. At the same time, it has become increasingly necessary to safeguard our facilities from attack. Information warfare is no longer the preserve of hackers and the like, but has matured to become an inescapable part of the planning and conduct of military operations.

Conclusion

Modern technology has made the Himalayas less of a challenge than they were earlier. The physical characteristics remain formidable but, to an increasing extent, means are becoming available to overcome them. Warfare is today far more clinical and we always have to be combat ready. Unfortunately for those that hold the purse-strings, whilst performance and capabilities of systems have increased and continue to improve, the requirements of the armed forces also increase. Continued research and development is both inescapable and inevitable. Yet, it must be emphasized that the capabilities that modern science has spawned are not available automatically to every country. They have to be developed or acquired. This is particularly true of India. Possibly, co-operative, coordinated and collaborative effort with other countries should be the option of choice for us.

Whilst the cost of preparing for war is increasing, the total cost of an actual war is far greater and its deleterious effects on military capability could last for an uncomfortably long time. Undoubtedly, if military capability is adequate, the adversary will be deterred but arguably, the most effective manner in which to bring about deterrence may be coercion and if necessary, information about the enemy. Intelligence is the key and money spent on intelligence gathering and analysis will always be money well spent.

History has shown that time and again, new military capabilities and equipment have been war-winning in their own right. This fact alone supports the recommendation for continued research and development of products that will improve surveillance and remote sensing. The nature of the Himalayan terrain makes the task far more difficult but equally, far more necessary.

Notes

Chapter 1

1 T. G. Montgomerie, Supt. G. T. Survey of India, Dehradun to the Lt. Governor, Punjab, 6 April 1871. See *Foreign Political A*. June 1871 (National Archives of India), pp. 560–97.
2 Alexander Cunningham, *Ladakh*, London (1854), p. 149.
3 R. L. Kennion to A.C. Talbot, Resident in Kashmir, 16 August 1900. See *Foreign Extl. B.* February 1901 (National Archives of India).
4 E. Joldan, *Harvest Festival of Buddhist Dards of Ladakh and Other Essays*, Srinagar (1985), pp. 64–5.
5 *Census of India, 1941, vol. XXII, Jammu and Kashmir* (1943), p. 481.
6 K. Warikoo, 'Ladakh's Trade Relations with Tibet under the Dogras', *China Report*, vol. 26, no. 2 (1990), p. 143.
7 F. Drew, *The Jummoo and Kashmir Territories*, London (1875), p. 363.
8 A. H. Francke, *Antiquities of Indian Tibet*, vol. 2 (1926), pp. 116–17.
9 Ibid., p. 40.
10 Ibid., pp. 55–6.
11 Sven Hedin, *Trans Himalaya*, vol. 1. London (1909), p. 57.
12 Ibid., p. 59.
13 Ibid., p. 58.
14 For further details, see note given by the Ministry of External Affairs, New Delhi, to the Embassy of China in India 24 September 1959. India, Ministry of External Affairs, *Notes, Memoranda and Letters Exchanged between the Governments of India and China, September to November 1959. White Paper II*, New Delhi (1959), pp. 88–96.
15 H. Strachey, *Account of Ladakh Trade, Foreign S.C.* 12 September 1852, pp. 153–6 (National Archives of India). For further details see K. Warikoo, op. cit., pp. 133–44.
16 Ibid.
17 Ibid.
18 Ibid.
19 Ibid.
20 Cunningham, op. cit., p. 251.
21 Moorcroft, *Travels*, London (1841), vol. 1, p. 253.
22 Moorcroft, who was in Leh at about the same time, was witness to the delayed arrival of a small caravan of 25 horses laden with shawl-wool, felts, tea and silk at Leh in the autumn of 1820 from the direction of Yarkand. See op. cit., p. 252.
23 C. B. Hugel, *Kashmir Under Maharaja Runjit Singh: Its Artistic Products, Taxation System, Imports, Exports and Trade*. Translated from original German and annotated by D. C. Sharma, New Delhi (1984), p. 61.

24 R. H. Davies, *Report on the Trade and Resources of the Countries on the North-Western Boundary of British India*, Lahore (1862), p. 6.
25 N. Elias, British Joint Commissioner, Ladakh to the Officer on Special Duty in Kashmir 12 July 1881. *Foreign Genl. B.* September 1881 (National Archives of India), pp. 144–5.
26 *Foreign Pol. A.* September 1875 (National Archives of India), pp. 21–4.
27 Ibid.
28 Ibid.
29 *Foreign Pol. A.* September 1878 (National Archives of India), pp. 45–69.
30 *Foreign Frontr. B.* February 1920 (National Archives of India), pp. 125–7.
31 *Foreign Pol. Frontr. B.* November 1921 (National Archives of India), pp. 16–18.
32 E. Joldan, op. cit., p. 54.
33 Janet Rizvi, *Ladakh: Crossroads of High Asia*, Delhi, 1983, p. 103.
34 E. G. Colvin, Resident in Kashmir to the Secretary to Government of India, Foreign Department dated 17 May 1904, *Foreign Frontr. A.* June 1904 (National Archives of India), pp. 110–11.
35 Ibid.
36 Ibid.
37 Cunningham, op. cit., p. 23.

Chapter 2

1 Russia occupied Kokand in 1876 and Britain went on another war with Afghanistan in 1878.
2 Cited in N. A. Khalfin, *Russia's Policy in Central Asia, 1857–1868*, translated from Russian into English by Hubert Evans, London (1864), p. 220.
3 Ibid.
4 Ibid., p. 221.
5 John Keay, *The Gilgit Game*, London: John Murray, p. 160.
6 Ibid., pp. 208–9.
7 D. Kaushik, *Central Asia in Modern Times*, Moscow (1970), p. 59.
8 See *Foreign Sec. I.* (1870), pp. 192–200.
9 See Memo of private interview between Lord Mayo and Maharaja Ranbir Singh at Sialkot on 3 May 1870.
10 Ibid.
11 This position was explained in much detail by Ranbir Singh in his memo dated 21 July 1870 discussing Kashmir's relations with Gilgit, etc.
12 G. J. Alder, *British India's Northern Frontier 1865–95: A Study in Imperial Policy*, London (1963), p. 11.
13 See memo of conversation held at Madhopore on 17 and 18 November 1876 between the Viceroy and Maharaja of Kashmir. *Foreign Secret* July 1877, 34–60B.
14 Ibid.
15 Lytton to Salisbury (Secretary of State) 11 June 1877, cited in ibid.
16 Ibid.
17 T. H. Thomton, Secretary, Foreign Department, GOI, to J. Biddulph, 22 September 1877. *Foreign Pol. A.* February (1878), pp. 117–37.
18 Leaving Gilgit on 17 May 1878, Shah Khushwakt took his route via Darel, Tangir, Maidan, Kabul, Mazar-i-Sharif, Bukhara, Samarkand, Khojend, Kokand, Marghilan, Osh, Andijan, Namangan, Karategin, Kolab, Wakhan, Hunza, Chitral, Darel and back to Gilgit. See *Foreign Secret* November 1879, pp. 152–91.
19 Ibid.
20 *Foreign Secret* January (1879), pp. 23–33.
21 Ibid.

22 Alder has aptly summed up the situation in these words: 'Amid a tangle of mountain tops higher than Mont Blanc and far from the reservoirs of British military strength, British influence at Gilgit was just not strong enough, either to impress Chiefs like Aman-ul-Mulk, or to control the dubious activities of the Kashmir officials.' See Alder, op. cit., p. 133.
23 Ibid., p. 138.
24 Secretary of State to GOI 16, September 1881. *Foreign Secret* January 1882, pp. 741–76.
25 King of Afghanistan to his Agent, General Amir Ahmed Khan, 20 March 1882, *Foreign Secret* April 1882, pp. 353–60.
26 In his reply to General Amir Ahmad dated 3 May 1882, the Secretary, Foreign Department GOI (C. Grant) made it explicitly clear that '[the] Government of India is under a solemn engagement to acknowledge the suzerainty of the Maharaja of Kashmir over Chitral and to afford His Highness countenance and material aid, if necessary in defending and maintaining his rights over that country'.
27 Panjdeh is famous for the 'Panjdeh crisis of 1985', when Britain and Russia came close to a war over a skirmish that occurred on 25 March 1985. The Russian forces seized the Afghan territory around an oasis at Panjdeh south of the Oxus River. However, the war was averted through diplomatic means after Russia kept the Merv oasis and relinquished other areas, besides promising to uphold the territorial integrity of Afghanistan. Following this incident, the Anglo-Russian Boundary Commission was established to delineate the northern frontier of Afghanistan and the border was finally laid down giving the border town of Kushka to Russia.
28 See D. K. Ghosh, *Kashmir in Transition, 1885–93*, Calcutta: World Press (1975), p. 159.
29 Cited in ibid., p. 160.
30 Government of India to the Secretary of State, 28 August 1885, *Foreign Sec. F.* December 1885, pp. 118–24.
31 G. J. Alder, op. cit., p. 155.
32 *Foreign Sec. F.* January 1888, pp. 115–18.
33 Ibid.
34 D. K. Ghosh, op. cit., p. 161.
35 *Foreign Sec. F.* January 1888, op. cit.
36 Cited in Ghosh, op. cit., p. 166.
37 H. M. Durand to the Resident in Kashmir, 18 October 1887, *Foreign Sec. F.* January 1888, op. cit.
38 Ibid.
39 Ibid.
40 Ibid.
41 H. M. Durand to Capt. A. G. A. Durand, 22 June 1888.
42 A. G. A. Durand, *Report on the Present Military Situation in Gilgit*, Simla, 5 December 1888, p. 14.
43 Ibid., p. 45.
44 Ibid., p. 14.
45 Ibid.
46 A. G. A. Durand, *Making of a Frontier*, London (1899), p. 120.
47 Ibid., p. 123. However, by virtue of a notification issued by the Foreign Department, Government of India, on 6 August 1889 Capt. Durand was appointed as the British Agent at Gilgit with effect from 17 July 1889.
48 See dispatch from Secretary of State to GOI 28 June 1889. In their letter of 6 May 1889 to the Secretary of State, the Government of India sought his approval to this proposal in view of the Russian advance 'up to the frontiers of Afghanistan

and recent development of her military power in Asia'. *Foreign Sec. F.* October 1889, pp. 104–32.

49 Capt. Durand identified the limits of Gilgit Agency with the Dardistan region, which comprised Chitral, Yasin, Punyal, Gilgit Valley, Hunza, Nagar, Astor Valley, Shin areas of Gor, Chilas, Darel, Tangir and Kohistan. See his *Making of a Frontier*, p. 198.

50 For further details see Capt. A. H. McMahon's (Political Agent in Gilgit) report to the Resident in Kashmir dated 10 May 1898. *Foreign Sec. F.* July 1898, p. 327.

51 Ibid.

52 This came to be in vogue a custom of exchange of gold dust and return presents between Hunza and China which continued to be followed until the early twentieth century except for the period when Yakub Beg ruled in Kashgharia.

53 As against his present of 15 miskalls of gold valued at about Rs. 120 at that time, the Hunza ruler used to receive return gifts of silk and cotton clothings, silver yamboo, tea-bricks etc. amounting to about Rs. 1,100.

54 While acknowledging such difficulties of the Hunza people, the Chinese *Taotai* in Kashgar granted them their right to cultivate lands in Raskam and also waived the land tax earlier proposed to be levied on them.

55 *Kashgar Diary*, 15 April 1898. *Foreign Sec. F.* June 1898, pp. 464–7.

56 Ibid. Macartney, the British Consul at Kashgar, learnt about this exchange of letters from his Chinese Munshi who had read the *Futai*'s reply addressed to Petrovsky.

57 *Kashghar Diary*, 15 January 1899. *Foreign Sec. F.* April 1899, pp. 119–25. See also *Foreign Sec. F.* May 1899, pp. 84–7.

58 *Foreign Sec. F.* August 1889, pp. 168–201.

59 Ibid. The *Amban* had in his letter dated the 7th day of the 3rd month of the 25th year of Kuangshu 1316, even proposed to supply grain annually to Hunza as would be sent to Yarkand for the purpose, taking due note of the food requirements of Hunza.

60 Ibid.

61 C. Scott to Salisbury, St Petersburg, 17 May 1899. See *Foreign Sec. F.* August 1899, p. 183.

62 Secretary of State to Viceroy, 1 July 1899. *Foreign Sec. F.* August 1899, pp. 168–201.

63 *Foreign Sec. F.* September 1899, pp. 210–40.

64 Secretary of State to the Viceroy, 1 July 1899.

65 *Foreign Sec. F.* September 1899, p. 216.

66 Ibid.

67 Bax-Ironside (from Peking) to the Foreign Office, London, 20 May 1899.

68 *Foreign Sec. F.* September 1899, op. cit.

69 *Foreign Sec. F.* November 1899, p. 25. See also telegram from the Governor of Sinkiang to G. Macartney, Kashgar, 8 November 1899. But this time only five plots were decided to be leased to Hunza, exclusive of those at Azghar and Ursur.

70 *Foreign Sec. F.* June 1900, pp. 83–95.

71 Ibid.

72 E. Satow (British Ambassador, Peking) to the Chinese Plenipotentiaries, 29 May 1901. *Foreign Sec. F.* July 1902, p. 56.

73 Alastair Lamb, *The Sino-Indian Border in Ladakh*, Canberra (1973), p. 59.

74 *Foreign. Sec. F.* September 1903, pp. 104–10.

75 Ibid.

76 Lamb has adroitly described the situation as: 'Petrovsky saw the Hunza move into Raskam as the thin edge of a British wedge. He was, in all probability, just as concerned at the prospect of the British turning the flank of the 1895 line in the Pamirs, as were the British at the prospect of the similar attempt by the Russians.' See Lamb, op. cit., p. 40.

77 *Foreign Sec. F.* August 1899, pp. 168–201. The Kashgar *Taotai*'s letter of 2 July 1899 addressed to the Mir of Hunza is indicative of the Chinese official opinion about the MacDonald proposals.
78 Calaude M. MacDonald in his dispatch dated 14 March 1899 to the Tsungli Yamen, Peking, had offered to relinquish India's claim to 'most of the Taghdumbash and Raskam districts' in exchange for the Chinese abandonment of 'her shadowy claim to suzerainty over the State of Kanjut' (Hunza).
79 In fact, Curzon was strongly in favour of implementing the MacDonald proposals of 14 March 1899 so that connection between Hunza and China could be severed once and for all. In doing so, Curzon was influenced by the fear of Russian domination of Chinese Turkestan which in turn would result in Russia staking a claim over Hunza, and also by the possibility of Safdar Ali's settlement in Raskam by the Chinese authorities in Kashghar, which would be a real source of discomfort to the new Mir of Hunza.
80 This was done by Lord Mayo in the course of his meeting with Ranbir Singh at Sialkot on 3 May 1870.
81 Lytton had detailed discussions with the Kashmir ruler on this issue at Madhopore on 17 and 18 November 1876. See *Foreign Secret* July 1871, pp. 34–60.
82 Mehtar of Chitral (Aman-ul-Mulk) to Viceroy dated 22 Zilhaj 1300 A.H. 22 September 1883. *Foreign Sec. E.* January 1884, pp. 124–51.
83 Khalfin, op. cit.
84 The British Embassy at St. Petersburg was officially informed by the Russian Government on 9(22) January 1909 that foreigners were not allowed to enter Russian possessions in Central Asia.
85 Galitzin had probably won Morimer's heart by being hospitable to a 'Britisher' named Beech at Ferghana in 1890. See R. Morimer to Viceroy, 23 May 1891. *Foreign Sec. F.* June 1891, pp. 94–98.
86 In his reply, dated 29 May 1891, to the British Ambassador at St. Petersburg, the Viceroy wrote: 'I wish he had not been invited'.
87 *Foreign Sec. F.* January 1893, p. 296.
88 *Foreign Sec. F.* October 1898, pp. 271–4. In his book published by the St. Petersburg Military District Headquarters in 1899, V. F. Novitsky discussed the British military arrangements in the north-western frontier of India.
89 *Kashgar Diary*, 16–31 October 1902. *Foreign Frontier B.* August 1903, p. 288.
90 *Foreign Sec. F.* November 1899, pp. 47–50.
91 Ibid.
92 M. De Stall to Lord Salisbury (British Foreign Secretary) 18 February 1899. *Foreign Sec. F.* May 1899, pp. 61–9.
93 Ibid. Salisbury to Viceroy of India, 14 March 1899.
94 Ibid. Viceroy to Salisbury, 17 March 1899.
95 Ibid.
96 Ibid.
97 Ibid. Salisbury to Viceroy, 15 April 1899.
98 Ibid. M. De Stall to Salisbury, 18 February 1899.
99 *Foreign Sec. F.* August 1907, pp. 41–67 A.
100 Ibid.
101 Ibid.
102 Ibid. Resident in Kashmir to Secretary, Foreign Dept. GOI, 21 June 1907.
103 *Ladakh Diary* 1 August 1907. *Foreign Frontier B.* October 1907, pp. 278–83.
104 Fielding to F. Younghusband, Resident in Kashmir, from camp Togru Su river, 5 August 1907. *Foreign Sec. F.* November 1907, pp. 1–4.
105 Ibid.
106 *Kashgar News Report*, 1–10 October 1907. *Foreign Sec. F.* January 1908, pp. 64–6.
107 Ibid. *Gilgit Diary*, 20–26 October 1907.

108 L. W. Dane, Secretary, Foreign Department GOI, to Richmond Ritchie, Secretary, Political Department, India Office, London, 9 January 1908. *Foreign Sec. F.* January 1908, pp. 64–6.
109 R. Ritchie to L.W. Dane, 21 February 1908. *Foreign Sec. F.* May 1908, p. 65.
110 Resident in Kashmir to Secretary, Foreign Department, GOI 30 April 1910. *Foreign Sec. F.* June 1910, pp. 3–5.
111 Ibid.
112 Ibid. See Lord Minto's minute dated 10 May 1910.
113 Ibid. See Secretary, Foreign Department GOI, to Resident 12 May 1910.
114 Resident to Secretary, Foreign Department GOI, 2 September 1910. See *Foreign Frontier B.* September 1910, p. 19.
115 Count Benckendorff to Sir Edward Grey, London, 28 November 1911. See *Foreign Sec. F.* September 1911, pp. 1–44.
116 Ibid.
117 Ibid. See Viceroy's telegram to the Secretary of State, 16 March 1912.
118 Ibid.
119 V. F. Novitsky *Voenniye Ocherki Indii*, St Petersburg (1899). Cited in Khalfin, op. cit., pp. 209–10.
120 Ibid.
121 Ibid.
122 Ibid.
123 See Komilov, *Otchetei o poezdkev Indiyu* (Report of Travels in India). Cited in Khalfin, op. cit., pp. 211–12.
124 Ibid., p. 194.
125 Ibid., pp. 194–7.
126 Ibid., p. 194.
127 In such a report addressed to the Director, First Department of Ministry of External Affairs dated 4(17) April 1907, Polotsov communicated details about the proceedings at the 22nd session of the Indian National Congress held at Calcutta in December 1906. See *Central State Historical Archives. Uzbek SSR. Tashkent*. Fond 1 C/2, Opis 2, Dela 301 (Microfilm, in National Archives of India).
128 Cited in Khalfin, op. cit., p. 216.
129 Ibid., p. 219.
130 This assurance was given by the Russian Ambassador at London, M. Sazanoff to the British Secretary of State, Lord Crewe on 29 September 1912. See *Foreign External B.* March 1913, pp. 54–62.
131 *Turkestanski Gazetti* 9(22) October 1903.
132 See *Foreign Frontier B.* May 1907, p. 262; November 1908, p. 98; August 1908, p. 615; February 1907, pp. 272–4.
133 R. P. Cobbold, *Innermost Asia*, London, 1900, p. 269.
134 Ibid.
135 Ibid.
136 *Foreign Frontier B.* February 1913, pp. 38–40.
137 *Foreign Frontier B.* June 1913, p. 47.

Chapter 3

1 Indian Ministry of External Affairs, *Reports of United Nations Commission for India and Pakistan* (June 1948 to December 1948), New Delhi, p. 18.
2 Ibid.

Chapter 5

1 A. Lamb, *Kashmir: A Disputed Legacy, 1846–1990*. Hertfordshire: Roxford Books, 1991, pp. 53–6.
2 Ibid., pp. 55–8.
3 Ibid., pp. 60–70.
4 Ibid., p. 72.
5 Dewan Pervez, *Jammu, Kashmir and Ladakh*. New Delhi: Manas Publications, 2004, pp. 129–31.
6 Ibid., pp 127–9.
7 Ibid., pp. 130–1.
8 Ibid., pp. 131–2.
9 Ibid., pp. 131–3.
10 Col. A. Durand, *The Making of a Frontier*. Karachi: Indus Publication 1977, p. 126.
11 Ibid.

Chapter 6

1 Ayaz, Amir, 'There is no Kashmir Solution'. *Dawn*, 9 December 2000.

Chapter 7

1 *The Daily Times*, 30 December 2004.
2 *The Daily Times*, 10 January 2005.

Chapter 8

1 *The Hindu*, 14 December 2003.
2 *Shivalik Chronicle*, Jammu, May 1992.
3 Ibid.
4 George Fernandes in an Interview with *Star News* on 18 July 1999.
5 See M. M. Khajooria, 'A Peep into the Ravaged Valley', *Kashmir Times*, 8 July 1997.
6 *Newsweek*, 7 June 2002.
7 Ibid.
8 See Dr. Siddiqui's write-up in *Dawn*, 28 May 2002.
9 Ibid.
10 See *Kashmir Times*, 9 October 2004.

Chapter 9

1 *Tibet File No. 2*, 'The Seventeen Point Agreement – May 1951' <http://www.freetibet.org/info/file/file2.html>; <http://www.tibetjustice.org/reports/sovereignty/independent/b/index.html>.
2 Michael Van Walt Van Praag (ed.), *The Legal Status of Tibet: Three Studies by Leading Jurists*, Central Tibetan Administration (CTA) 1989. UN General Assembly Resolutions reprinted by CTA 'International Resolutions and Recognitions on Tibet (1959–97)' (1997), pp. 4–6. Also see *Tibet: Proving Truth from Facts*, CTA (1996).
3 Tashi Rabgey and Tseten Wangchuk Sharlho, 'Sino-Tibetan Dialogue in the Post-Mao Era: Lessons and Prospects' *Policy Studies no. 12*, Washington: East West Centre (2004).

4 Julian Gearing, 'Tibet: Between Surrender and Struggle', *Asia Times Online Ltd.* Full text on 'Regional Ethnic Autonomy in Tibet' can be seen on *Beijing Time* or *China Daily*, 24 May 2004.

5 P. Stobdan, 'Democracy and Future of Tibet', *Hindustan Times*, 16 April 1996.

6 Cristina Rocca quoted Martin Baumann, a German researcher, having suggested in 1997 that there were three to four million Buddhists in the US and there was a rise of 80 per cent between the 1996 and 2001 censuses in Australia, 'Westerners Associate Buddhism with Therapy' – interview by Cristina Rocca with Avijit Gosh, *Times of India*, 16 May 2006.

7 For full text of the proposal, see <http://www.dalailama.com/page.96.htm>.

8 The comments were made by Xiangba Pingcuo, who is also deputy Communist Party Secretary of Tibet, on the sidelines of China's annual Parliament session in March 2006.

9 Said by Lodi Gyari following the talks in Beijing, *Reuters*, 6 March 2006. Also see the statement of the Kashag on the 46th Anniversary of the Tibetan People's Uprising Day, issued on 10 March 2005.

10 Nicholas Haysom played the key role in negotiating the Arusha peace deal in Burundi (2000) and the Naivasha Accord for Sudan.

11 'Dalai Lama Renews Pledge for Tibet to Stay Within China', *AFP*, 10 March 2005. Also see Dalai Lama's 46th anniversary speech in 2005.

12 'Tibet is backward, it's a big land, quite rich in natural resources, but we completely lack the technology or expertise (to exploit them). So if we remain within China, we may get a greater benefit, provided it respects our culture and beautiful environment and gives us some kind of guarantee . . . for us, [it would mean] more modernization,' he said in an interview to *Time* magazine as quoted by PTI and the *Hindu*, 18 October 2004. In the past, he dreamt about finding some way of combining Marxism and Buddhism to a genuine socialist society through the development of altruism.

13 In a statement, he said: 'I have expressed a desire to make a trip to China because it is a country of ancient Buddhist traditions, where many important, sacred shrines are found, it is a place of pilgrimage which I would be happy to visit.' He added: 'I hope to be able to see for myself the changes and developments in the People's Republic of China.' For the Dalai Lama's 10th March statement, see the Tibetan Government in Exile website, <http://www.tibet.com>.

14 'China Open to Visit by Dalai Lama', *Hindustan Times*, 4 April 2006. Also read 'Toe our line and be back China to Dalai Lama' – interview by Chinese Ambassador to India, Sun Yuxi, *Hindustan Times*, 17 February 2006.

15 For quite some time, the Dalai Lama has been expressing his desire to meet Chinese leaders to discuss the issue of Tibet directly with them.

16 In recent years, hundreds of Chinese visited Dharamsala and other places in India, to meet the Dalai Lama and for taking religious vows from him.

17 'China Hosts World First Buddhist Forum', *CCTV*, 13 April 2006. Luis Ramirez, 'China World Buddhist Forum Excludes Dalai Lama', *Xinhua*, 13 April 2006.

18 'Tibet Campaign May Boil Over After Dalai Lama', *Reuters*, 1 January 2006.

19 P. G. Rajamohan, 'Terrorist Attack in Bangalore: A Profile', *IPCS Special Report 10*, January 2006, <http://www.ipcs.org/IPCS-Special-Report-10.pdf>

20 'Tibet Protest Was Not Directed Against Institute: IISc', *Hindu*, 31 December 2005.

21 Pankaj Mishra, 'The Restless Children of the Dalai Lama', *New York Times*, 18 December 2005.

22 Discussion with Chaman Sharma, an activist for the Tibet cause, who is fully committed to a violent struggle.

23 Masood Butt, *Dalai Lama and the Muslims of Tibet*, Office of Tibet, New York, 19 February 2005.

24 Pankaj Mishra, op. cit.

25 The statement of the Kashag on the 46th Anniversary of the Tibetan People's Uprising Day, issued on 10 March 2005 reflects the internal problems and contradictions on the issue of the seemingly fast track approach of the Dalai Lama's team with Beijing and the manner in which his administration is trying to bring about a general consensus on the approach.

26 Anand S. T. Das, 'China's attitude towards Tibet changing: PM-in-exile', *Indian Express*, 3 May 2006.

27 China asked foreign governments to be on 'high alert' against the Dalai Lama's political activities and bid for 'Tibetan independence' on the 'pretext of religion', *PTI*, 12 May 2006.

28 Subramanian Swamy, 'Sino-Indian Relations Through the Tibet Prism', *Frontline*, 17(18), 2–15 September 2000.

29 A Declaration on Principles for Relations and Comprehensive Cooperation between India and China signed on 23 June 2003 contains India's changed formulation on Tibet which says 'the Tibet Autonomous Region is part of the territory of the People's Republic of China' and reiterates that it does not allow Tibetans to engage in anti-Chinese political activities in India. India's Foreign Secretary reiterated India's Tibet policy and said 'Tibet is an autonomous region of the People's Republic of China. But we respect the Dalai Lama as a very important religious figure and a very highly respected figure.' *No Change in India's Tibet Policy*, Shyam Saran, *PTI*, 31 March 2006.

30 'Rahul Seeks Dalai's Blessings', *Hindustan Times*, 24 May 2006. Rahul Gandhi extended an invitation to the Dalai Lama to visit his family in New Delhi.

31 India's foremost strategic commentator and former National Security Advisor, J. N. Dixit said 'China remains seriously concerned about the activities of the Dalai Lama with his base in India. We must ensure that Tibetan ethno-religious politics does not affect our interests in terms of maintaining stability in our relations with China.' See 'Tibetans must not affect our relations with China', <http://www.rediff.com/news/1996/2511chi2.htm>.

32 'Indian Foreign Secretary visits Dharamshala', *Phayul*, 18 March 2006. Shyam Saran was in Dharamshala as a part of protocol to meet the Dalai Lama. He also met officials of the Tibetan exile Government, including the Tibetan Prime Minister Prof. Samdhong Rinpoche and the Karmapa. Though the meeting with the Dalai Lama was routine in nature, the media linked it to the visit by the State Councilor of China, Tang Jiaxuan to Kathmandu during the same week as a move to counter Chinese diplomatic manoeuvering in Nepal.

33 Perception gathered after the author's discussion with the Dalai Lama's elder brother Gyalo Thundup in Leh, August 2005.

34 Subramanian Swamy, 'Sino-Indian relations through the Tibet Prism', *Frontline*, 17(18), 2–15 September 2000.

35 In 1947, the Tibetan authorities in Lhasa wrote to the Government of India to return territories from Indian border areas extending from Bhutan to Ladakh, including Darjeeling and Sikkim. Such a claim of territory from India by the Tibetan government was kept confidential for 12 years. It was only in 1959, when Nehru revealed it in the Lok Sabha that it became public knowledge. He admitted, 'It is true that we received a telegram from the Tibetan Bureau in Lhasa, claiming the return of Tibetan territory on the boundary of India and Tibet' Read Yang Yun-Yuan, 'Controversies over Tibet: China versus India, 1947–1949', *China Quarterly*, no. III, September 1987, p. 410.

36 This point is articulated by Chinese academics. Zhao Gancheng, Head of South Asia Studies, Shanghai Institute of International Studies, said during his visit to New Delhi in 2003, that Tawang is the birthplace of the sixth Dalai Lama and its surrender to India would hurt the sentiments of the Tibetan people. The

Chinese Institute of International Studies (CIIS), in its publication *International Studies Journal*, mentioned in 1993 for the first time about Monyul in Tawang and its importance as the birthplace of the sixth Dalai Lama.

37 Rajesh S. Kharat, *Tibetan Refugees in India*, Delhi (2003).

38 There is normally a floating population of around 500–600 unregistered Tibetan refugees at any time in and around Dharamshala.

39 The author witnessed the Dalai Lama's *Kalachakra Tantra* (5–16 January 2006) in Amaravati, Andhra Pradesh and collected data first hand.

40 Interview with several Tibetans residing in Majnu ka Tila in Delhi suggests that a much wider web of trans-border links exist and the Indian agencies have very little clue about this.

41 During a meeting at the Tibetan Reception Centre in Chengdu in 1992 the author was introduced to several Tibetans who had served in the Special Frontier Force (SFS).

42 The notorious wildlife trader Tsewang was finally arrested in Nepal. The CBI had been looking for Tsewang and Tashi Tsering, South Asia's most wanted wildlife criminals and wanted in several cases in India. They were arrested on information provided by field operatives of Wildlife Conservation Nepal (WCN) and the Wildlife Trust of India (WTI). In February 2006, Rajasthan police arrested Tibetan Neema Kampa, the gang leader who smuggled animal pelts out of India. Poacher Sansar Chand, too, revealed that the hundreds of leopard and tiger skins he sold to international dealers, mostly from Nepal, passed through Tibet. Chand was arrested on 30 June 2005. His interrogation also revealed the widespread network and the route of the international wildlife trade. Chand named his prominent Tibetan clients in 'Poacher Sansar Chand Revealed Tibet, Nepal Links', CBI <http://www.wpsi-india.org/news/06022006.php>. See 'Notorious Wildlife Trader Tsewang Arrested in Nepal', *Indian Express*,' 14 December 2005; 'Is Tsewang none other than South Asia's Most-Wanted Wildlife Criminal Tashi Tsering?', *Indian Express*, 6 February 2006.

43 She told *India TV* in a panel discussion on wildlife poaching in India and its reported smuggling to Tibet. See 'Indian Tiger Skins Flooding Tibet Black-Market', *Indian Express*, 7 September 2005.

44 'Dalai Lama Attempts to Calm Storm in India on his Kashmir Comments', *AFP*, 7 August 2001.

45 Mirwaiz Umer Farooq, Hurriyat leader said, 'The Dalai Lama has given a wonderful statement' <http://www.rediff.com>. Mirwaiz said that the Hurriyat leadership would try to meet the Dalai Lama: 'If we are not able to meet him then we will send him a letter. I have already extended him an invitation to visit Kashmir. We will send him an invitation in writing and I am sure he will respond to it.'

46 'No Question Middle Path on Kashmir: BJP', *The Times of India*, 8 August 2001.

47 'India Seeks Full Report on Dalai Lama's Kashmir Statement' and 'Govt Miffed by Dalai Lama's Statement on J&K' <http://www.rediff.com/news/2001/aug/06onkar.htm>

48 'Emissary denies Dalai Lama called for self-rule in Kashmir', *Indo-Asian News Service*, 6 August 2001.

49 'Tibetans Dismayed by Attack on Dalai Lama's Vehicle', *Indo-Asian News Service*, 9 August 2001.

50 'Dalai Lama Encourages Hindu-Muslim Dialogue Over Ayodhya Dispute', *The Times of India*, 14 January 2004.

51 'Kanchi Seer Questions Dalai Lama's Offers of Mediation in Ayodhya Talks', *The Times of India*, 2 February 2004.

52 'Should "Guest" Dalai Lama interfere on Babari?' *The Milli Gazette*, <http://www.milligazette.com/Archives/2004/01-15Feb04-Print- Edition/011502200493.htm>.

53 P. Stobdan, 'Tibet and the Himalayas: A Strategic Dimension to India's Security', *Strategic Analysis*, XIV(6), 6 September 1991, pp. 687–702.

54 Tenzin Gyatsho, Dalai Lama XIV, *Freedom in Exile: The Autobiography of His Holiness the Dalai Lama of Tibet*, London: Hodder & Stoughton, 1990, pp. 245, 257. *Tibet: From Buddhist Polity to Invasion and Diaspora*, Australia: Department of International Relations, FHSS, Bond University (2005).

55 The Chinese have been insistent in pushing Gyaincain Norbu as their choice for the new Panchen Lama, as against Dalai Lama's chosen boy Gedhun Choekyi Nyima (1995).

56 P. Stobdan, 'Panchen Lama: China's Tibet Policy', *Hindustan Times*, 31 December 1995.

57 This impression was gathered during the author's interactions with Chinese officials and academia in Sichuan, Beijing and Yunnan in 1992 and 1997.

58 In 1992, the Chinese enthroned Urgyen Thinley Dorje as the seventeenth Karmapa. Dorje was recognized by the Dalai Lama as well. He fled from Tibet to India in January 2000, supposedly in connivance with Chinese agencies, whose main goal in India was to gain control over Rumtek Monastery in Sikkim and the famous Black Hat of Karmpa Kagyu order.

59 Dr. Saklani, in his study on the ideological trends among the Tibetan refugees in India, remarked that the overwhelming articulation of their self-chosen travails of refugeehood were in religious rather than political terms.

60 According to the Dalai Lama administration, the Tibetan refugees have constructed 181 monasteries and eight nunneries in India, Nepal and Bhutan since 1959, representing all five traditions of Tibetan religion. They have come up in places like Shimla, Manali, Dehradun, Kalimpong, Darjeeling, Leh, Gangtok, Palampur, Bejnath, Bir, Dharamsala and others. They house 17,376 monks and 549 nuns. The Dalai Lama's Department of Religion and Culture claimed that these monasteries function under its supervision.

61 A publication by Department of Religion and Culture, Central Tibetan Administration of the Dalai Lama, *gTor-bSig-tang-Gsar- bSkrn-gi-B-tSl* (Destruction and Reconstruction) 2000, is devoted to rebuilding of Tibetan monasteries in various parts of India.

62 See Integrated Development Plan (IDP) of the Tibetan Government in Exile, <http://www.tibet.net/publication/pc/download/IDP3%20rationale.pdf>.

63 The Dorje Shugden cult – the spirit deity which originated in Tibet in the seventeenth century, coinciding with the fifth Dalai Lama – has been a source of conflict in the Tibetan community both living in India and abroad. It was in 1996 that the Dalai Lama persuaded the Tibetans and Buddhists to stop worshipping the Shugden deity, with some success, though some Tibetan groups have refused to accept his dictum. The Shugden devotees murdered a high Lama and his disciples in Dharmshala in February 1997, which caused much tension in the Tibetan community. This led the Indian Government to put some Shugden activists living in India under observation. While the Tibetans generally support the Dalai Lama on his stand over Shugden, China is reported to be supporting the Shugden cult in Tibet. More recently, in 2007, the Dalai Lama asked the Tibetans to stop worshipping Shugden, which he described as not being a Buddhist practice and contrary to the unity of Tibetans.

64 Though the Karmapa lineage (thirteenth century) pre-dates the Dalai Lama lineage, it has not played any major role in the politics of Tibet. The controversy over the seventeenth Karmapa hinges around the existence of two Urgyen Trinley Dorje (UTD) (based around the existence of two Karmapas in India: (i) near Dharmshala, who was recognized by Tai Situ Rinpoche in 1992 as the Karmapa, and (ii) Thinley Thaye Dorje (based in Delhi) who was recognized by Shamar Rinpoche as him. The UTD has received the support and

recognition of both China and the Dalai Lama, which has been a cause of concern in India.

65 The Government of India announced it was to include Bhoti (Tibetan) language under Schedule VIII of the Indian Constitution in May 2006.

66 See for details, Tanka B. Subba, 'Tibetans in Exile: Economic Pursuits and Ethnicity,' in K. Warikoo and Dawa Norbu (eds) *Ethnicity and Politics in Central Asia*, New Delhi: South Asian Publishers (1992), pp. 204–19.

67 The Dalai Lama made it clear to the Chief Minister of Himachal Pradesh that 'he would reconsider his decision of not moving his headquarters from Dharmshala only if there was an assurance from senior local residents that the incidents that occurred there recently would not be repeated.' 'Dalai Lama hints at reconsidering move', *The Times of India*, New Delhi, 12 May 1994. 'Indian Official Asks Dalai Lama to Stay' <http://www.tibet.ca/en/wtnarchive/1994/5/12-4_3.html>.

68 Tibetans acquired landed property allegedly under pseudo names in places like Rohru, Hatkoti, Khara Pathar, Hunur, and Kotkhai in Himachal Pradesh. See J. N. Sadhu, 'Tibetan Refugees in HP Acquiring Property', *Indian Express*, 15 September 1983. Also cited by Tanka B. Subba, op. cit., p. 12.

69 P. Stobdan, 'Destabilising Presence of Tibetans', *Indian Express*, 8 June 1994.

70 Although, in order to pre-empt the Chinese from gaining control over his successor, the Dalai Lama has on one occasion said that he will be reborn among the Tibetan community in exile. In a recent interview, he said that 'he could die happily in India if Tibet's political status did not change in his lifetime'.

71 Tibetans in India are being used by foreign agencies and missions for information. In the past, Human Rights activists used the Tibetan Muslims to ascertain ground positions in Kashmir.

72 China lays claim to 90,000 km^2 of land in Arunachal Pradesh and considers it as a part of the Cona County Tibet Autonomous Region. China does not recognize the McMahon Line, and instead calls the area South Tibet (pinyin: Zàngnán), dividing it into six border counties of TAR: Cona, Lhünzê, Nang, Mainling, Mêdog, and Zayü. On 25 July 2003, Chinese Foreign Ministry spokesman Kong Quan said China does not recognize that Arunachal Pradesh is part of India. As late as 4 April 2005 China's envoy to New Delhi, Sun Yuxi, said ahead of the landmark visit by Chinese Premier Wen Jiabao that Arunachal Pradesh was still 'a disputed area' between India and China.

Chapter 10

1 Asad Husain, *British India's Relations with the Kingdom of Nepal* (London: George Allen and Unwin, 1970), p. 24.

2 For an historical account of some of these aspects the following studies may be useful: D. R. Regmi, *Ancient Nepal* (Firma K. L. Mukhopadhyay: Calcutta, 1960); Toni Hagen, *Nepal, the Kingdom in the Himalayas* (Berne: Kummerly and Frey, 1961); Francis Hamiltton, *An account of the Kingdom of Nepal* (Edinburgh: Archbishop Constable & Co., 1819); R. C. Mazumdar, *The History and Culture of the Indian People, The Classical Age*, Vol. III (Bombay: Bharatiya Vidya Bhawan, 1962).

3 Vidya Bir Singh Kansakar, 'Nepal–India Open Border: Prospects, Problems and Challenges', at <http:www.ifa.org.np/talk/trade.php> (accessed 30 October 2006).

4 C. U. Atchison, *A Collection of Treaties, Engagements, Sanads Relating to India and Neighbouring Countries*, Vol. II (Calcutta: Bengal Printing Company, 1863).

5 Ludwig F. Steller, 'Modern Nepal' in K. P. Malla (ed.) *Nepal: Perspectives on Continuity and Change* (Kathmandu: CNAS, 1989).

6 Asad Husain, op. cit.
7 Sushila Tyagi, *Indo-Nepalese Relations, 1858–1914* (Delhi: D. K. Publishing House, 1974), pp. 87–8.
8 Nepal did not conclude any formal Treaty with Sikkim but returned its occupied territory under the Treaty of Segauli itself. It was recognized in the Treaty of Titaliya concluded by the East Indian Company with the Raja of Sikkim on 10 February 1871.
9 Kansakar, op. cit.; V. B. S. Kansakar, 'Nepal–India open Border: Nature, Pattern and Cultural Implications' in Ramakant B. C. Upreti (eds., *India and Nepal: Aspect of Interdependent Relations* (Delhi: Kalinga Publications, 2001), pp. 1–14.
10 P. P. Karan and W. Jenkins, *Nepal: A Physical and Cultural Geography* (Lexington: University of Kentucky Press, 1960), Harka Bahadur Gurung, 'Geographic Foundations of Nepal', *The Himalayan Review*, vol. I, 1968.
11 Hari Bansa Jha, *Nepal Me Terai Samudaya Ebam Rastriya Aekta* (The Terai Community and National Integration in Nepal) (Kathmandu: CETS, 1994), p. 17.
12 B. C. Upreti, *Marginal Migrants*, Delhi: Kalinga Publications, 2002.
13 Joseph W. Edden, *Planned Resettlement in Tarai* (Kathmandu: INAS & CEDA, 1974); Sohan Ram Yadav, *Rural and Agrarian Social Structure of Nepal* (New Delhi: Commonwealth Publications, 1992).
14 Frederick H. Gaige, *Regionalism and National Unity in Nepal* (New Delhi: Vikas Publishing House, 1975).
15 Kansakar, op. cit.
16 Ashok Mehta, 'Problem of Terrorism and other Illegal Activities on Indo-Nepal Border: Issues in Effective Border Management' in Ramakant and B. C. Upreti (eds.), op. cit., p. 20.
17 *The Kathmandu Post*, 28 February 2003.
18 *POT Nepal Series* (New Delhi), 3 January 2003, p. 4.
19 *The Kathmandu Post*, 3 June 2003.
20 S. D. Muni, *India and Nepal: A Changing Relationship* (New Delhi: Konark Publishing Ltd, 1992), p. 129.
21 Ibid.
22 Ram Prasad Rajbahak, *Nepal–India Open Border: A Bond of Shared Aspirations* (New Delhi: Lancer, 1992), p. 11.
23 *Kathmandu Post*, 27 March 2003.
24 *Kathmandu Post*, 30 November 2003.
25 *Kathmandu Post*, 16 March 2004.
26 Fredrick H. Gaige, op. cit., p. 50.
27 Myron Weiner, 'On International Migration and International Relations', *Population and Development Review*, vol. 1, no. 3, September 1985, p. 444.
28 Lok Raj Baral, *Regional Migration, Ethnicity and Security* (New Delhi: Sterling Publishers, 1990), p. 114.
29 Lok Raj Baral, 'Political Demography and Security in South Asia' in Sridhar K. Khatri (ed.), *Regional Security in South Asia* (Kathmandu: CNAS, 1987), p. 151.
30 V. B. S. Kansakar, 'Population in Nepal', in Kamal P. Malla (ed.), *Nepal: Perspectives on Continuity and Change* (Kathmandu: CNAS, 1989), p. 37.
31 The Nepalese Government has tried to discourage large-scale migration by adopting new policies concerning citizenship, industry and labour. In 1984 a task force on migration was constituted to study the consequences of immigration and to suggest remedies, but no action was taken on the report produced.
32 Piers Blaikie *et al.*, *Nepal in Crisis, Growth and Stagnation at the Periphery* (Oxford: Oxford University Press, 1980), p. 19.
33 Ibid.
34 Fredrick H. Gaige, op. cit., p. 202.

35 Yugeshwar P. Varma, *Nepal: Progress and Problems (1972–80)* (Lalitpur, Nepal: Pratibha Verma, 1981, p. 168); Ramakant and B. C. Upreti, 'Regionalisation in Nepal: A Study of the Tarai Region' in Ramakant *et al.*, *Regionalism in South Asia* (Jaipur: Aalekh, 1984); Ramakant and B. C. Upreti, 'Regionalism in Nepal' in Urmila Phadnis *et al.*, *Domestic Conflicts in South Asia, vol. II* (New Delhi: South Asian Publishers, 1986).
36 Lok Raj Baral, op. cit., p. 153.
37 *Kathmandu Post*, 5 July 2003.
38 *Spotlight* (Kathmandu), 25–31 July 2003.
39 *Kathmandu Post*, 30 November 2003.

Chapter 11

1 J. C. White, *Sikkim and Bhutan: Twenty-One Years on the North-East Frontier – 1887–1908* (London: 1909), pp. 264–5; A.R. Field claims that British had relations with Bhutan long before in 1767. See A. R. Field, 'A Note Concerning Early Anglo-Bhutanese Relations', *East and West* (Rome) (1960), pp. 340–5.
2 Kapileshwar Labh, *India and Bhutan*, New Delhi (1974), p. 173.
3 Manorama Kohli, *India and Bhutan: A Study in Inter-Relations 1772–1910*, New Delhi (1982), pp. 155–185.
4 Kapileshwar Labh, op. cit, pp. 181–3.
5 Ram Rahul, *Royal Bhutan*, New Delhi (1983), p. 29.
6 Kapileshwar Labh, op. cit., p. 183.
7 V. H. Coelho, *Sikkim and Bhutan*, New Delhi (1971), p. 71.
8 Ibid.
9 Kapileshwar Labh, op. cit., p. 209.
10 Nirmala Das, *The Dragon Country*, Bombay (1974), pp. 48–9.
11 *Foreign Policy of India, Text of Documents: 1947–59*, New Delhi, December 1959, pp. 15–20.
12 Nagendra Singh, *Bhutan: A Kingdom in the Himalayas*, New Delhi (1972), pp. 131–2.
13 *Selected Works of Mao Tse Tung*, Vol. III, Bombay (1954) p. 72.
14 S. Gopal, *Jawaharlal Nehru: A Biography*, Vol. Two, 1947–56, Delhi (1984), p. 176.
15 Lt. Gen. B. M. Kaul, op. cit., p. 248.
16 *Foreign Policy of India, Text of Documents, 1967–74*, New Delhi (1964), p. 8.
17 Leo E. Rose, *The Politics of Bhutan*, Ithaca and London (1974), p. 74. Also see Leo E. Rose, 'Bhutan's External Relations' *Pacific Affairs*, 47(2), Summer 1974, p. 194.
18 Ibid., p. 74.
19 A. Appadorai, *Selected Documents on Indian Foreign Policy, 1947–72, Vol. I*, Delhi, (1982), p. 548.
20 Kapileshwar Labh, op. cit., p. 214.
21 Ibid.
22 L. S. Rathore, *The Changing Bhutan*, New Delhi (1974), p. 79.
23 K. Krishna Moorthy, 'Bhutan, Thoughts of Sovereignty,' *Far Eastern Economic Review* (Hong Kong), 16 February 1961, p. 297.
24 K. S. Mathew, 'Bhutan', *The Illustrated Weekly of India* (Bombay), 1 July 1973, p. 39.
25 As told by the then Indian Ambassador to Bhutan, Ashok Gokhale (1972–74), and *The Stateman*, 13 November 1974.
26 C. K. Tiwari, 'India's Northern Security and the Himalayan State', *Asian Profile*, 14(5), August 1986, p. 445.
27 Rose, op. cit., p. 81.

28 Klaus H. Pringsheim, 'China, India and their Himalayan Border (1961–63)', *Asian Survey*, 3(10), October 1963, p. 486.
29 *The Motherland* (New Delhi), 25 December 1974.
30 Neville Maxwell, *India's China War*, Bombay (1971), p. 354.
31 Ibid.
32 Rose, op. cit., p. 82.
33 Ibid.
34 Kapileshwar Labh, op. cit., p. 214.
35 Ram Rahul, op. cit., p. 39.
36 Nari Rustomji, *Bhutan: The Dragon Kingdom in Crisis*, Delhi (1978).
37 It was reported that after the assassination, civil war in Bhutan broke out between Dorji's and Wangchuk's family. Dorji's anti-king groups wanted to modify the Royal order and Bhutan's relations with India. For a detailed report, see Madhusudan Bhattacharya, 'Bhutan in Turmoil', *Mainstream*, 19 December 1964, p. 8.
38 Rose, op. cit., pp. 85–6.
39 Nari Rustomji, op. cit.
40 Ibid.
41 P. S. Bhati, 'Foreign Policy of Bhutan', *The Indian Journal of Political Studies*, 11 December 1987, p. 99.
42 Ibid.
43 Rose, op. cit., p. 87.
44 Dr. Zakir Hussain, 'Speeches on Bhutan', *Foreign Affairs Record*, New Delhi, MEA Publications (1968), pp. 27–9.
45 B. S. Das was appointed as Special Officer to Thimpu, 1968–72. For details, see B. S. Das, *Mission to Bhutan: A Nation in Transition*, New Delhi (1995).
46 Ibid.
47 Ibid.
48 From the Indian side, B. S. Das was appointed as first Indian Ambassador to Bhutan on 15 July 1971. As told by B. S. Das in an interview.
49 Valentine J. Belfigilo, 'India's Economic and Political Relations with Bhutan', *Asian Survey*, August 1972, p. 685.
50 An interview with D. S. Soman (IPS), Internal Security Advisor to the King of Bhutan, from 1969–74, who was eye-witness to many important events in Bhutan. Also see *The New York Times*, 3 June 1974.
51 Rose, op. cit., p. 103.
52 Ibid.
53 Das, op. cit., p. 122.
54 Jigme Singye Wangchuk said this at an interview with *Asia Week*, 27 September 1985, p. 47.
55 Rose, op. cit., p. 104.
56 *The New York Times*, 7 June 1974.
57 *The Statesman*, 13 November 1974.
58 Ibid.
59 *The Times of India*, 23 December 1974.
60 Ibid.
61 *Hindustan Times*, 14 August 1976.
62 S. D. Muni, 'India's Beneficial Bilateralism in South Asia', *India Quarterly*, 25(4), October–December 1979, pp. 417–33.
63 Ibid, p. 418.
64 *Deccan Chronicle*, 28 August 1977.
65 Ibid.
66 *Foreign Affairs Record*, 22(11), November 1977, p. 219; *The Hindu*, 23 November 1977.
67 *The Hindu*, 23 November 1977.

68 *The Hindustan Times*, 27 March 1978.
69 Ibid. Also see the text of *King of Bhutan's Speech*.
70 *The Hindustan Times*, 25 March 1978.
71 *National Herald*, 24 November 1978.
72 Mohan Ram, 'Bhutan: Keeping a Neighbour in Line', *Far Eastern Economic Review*, 30 May 1980, p. 32.
73 An interview with *Asia Week*, 27 September 1985, p. 47.
74 *The Times of India*, 11 September 1979.
75 *The Patriot*, 14 September 1979.
76 *New Wave*, 28 March 1982 and *The Indian Express*, 11 November 1982.
77 Ibid.
78 *The Times of India*, no. 110.
79 Ibid.
80 Manorama Kohli, 'Bhutan's Strategic Environment: Changing Perception', *India Quarterly*, 42(2), April–June 1986, p. 149.
81 S. D. Muni, 'Sino-Bhutanese Boundary Talks', *The Indian Express*, 12 June 1984.
82 S. D. Muni, op. cit., pp. 515–16.
83 Rathore, op. cit., pp. 20–21.
84 S. D. Muni, 'Sino-Bhutanese Boundary Talks', *The Indian Express*, 12 June 1984.
85 *The Hindustan Times*, 17 December 1996.
86 Ibid.
87 S. D. Muni, 'Bhutan Steps out', *The World Today*, 40(12), December 1984, p. 516.
88 Ibid.
89 S. D. Muni, 'Sino-Bhutanese Boundary and Implications for India', *Mainstream*, 24(41), 14 June 1986, pp. 17–20.
90 Ibid.
91 S. D. Muni, op. cit.
92 *The Kuensel*, 2 May 1985.
93 Mohan Ram, 'Bhutan: Solution at Sight', *Far Eastern Economic Review*, 10 July 1986, p. 28.
94 *The Times of India*, 15 May 1988.
95 *The Kuensel*, 4 June 1988.
96 Ibid.
97 *The Times of India*, 15 May 1988.
98 Ibid.
99 *The Kuensel*, 1 September 1990.
100 Ibid.
101 Ibid.
102 Ibid.
103 'National Assembly Debates', *The Kuensel*, 10 August 1996, p. 17.
104 'National Assembly Debates', *The Kuensel*, 9 August 1997, p. 2.
105 Ibid.
106 Ibid.
107 'Bhutan–China Border Talks', *The People's Review*, 24–31 December 1998. <www.yomari.com/p-review/1998/12/241298/bhut.html>.
108 Ibid.
109 *The Kuensel*, 25 September, 1 October 1999, p. 1.
110 *The Kuensel*, 27 July 2001, p. 1.
111 <www.yomari.com/p-review/1998/12/241298/bhut.html>.
112 Ibid.
113 Ibid.
114 Ibid.

115 'National Assembly Debates', *The Kuensel*, 9 August 1997, p. 2.
116 Ibid.
117 *The Kuensel*, 16 September 1995.
118 *The Kuensel*, 9 August 1997.
119 Rose, op. cit., p. 96.
120 Ibid.
121 *The Kuensel*, 4 August 1990.
122 *The Hindu*, 28 October 1990; *The Kuensel*, 23 February, 1991.
123 Parmanand, *The Politics of Bhutan: Retrospect and Prospect?*, New Delhi: Pragati, (1992), p. 145.
124 S. K. Pradhan, 'Human Rights Situation in Bhutan, Appeal by PFHRB', *IMADR Review for Research and Action*, 4, September 1994, p. 3.
125 *The Kuensel*, 27 May–2 June 2000.
126 John Quigley, p. 190.
127 15th Ministerial Joint Committee Press Release, Kathmandu: Nepal Ministry of Foreign Affairs, 22 October 2003.
128 During the process of verification an angry mob attacked the Bhutanese officials ending the break-up of the meeting on 22 December 2003.
129 'US Offer to Take Refugees Could Break Bhutan-Nepal Deadlock', US Department of State, at <http://usinfo.state.gov>, 6 October 2006.
130 Ibid.
131 Ibid.
132 'Refugee Leaders not Happy with US Settlement Offer', *Rising Nepal*, 6 November 2006, http://www.gorkhapatra.org.np/>.
133 'Third Country Settlement Optional: UNHCR Official', *Rising Nepal*, 11 November 2006, <http://www.gorkhapatra.org.np/>.
134 Ibid.
135 'Solution in Sight?' *The Kuensel*, 11 October 2006, <www.Kuenselonline.htm>.
136 *The Kuensel*, 25 June 2000.
137 *Bhutan: South Asia* Assessment (2003), <http://www.satp.org/satporgtp/south asia/index.html>.
138 Jaideep Saikia, *Terrorism Sans Frontiers: ULFA Digs Deeper into Bhutan*, in O. P. Misra and Sucheta Ghosh (eds.) *Terrorism and Low Intensity Conflict in South Asian Region*, New Delhi: Manak (2003), p. 447.
139 Ibid.
140 Ibid.
141 Ibid., p. 452.
142 *The Kuensel*, 20 September 2003.
143 Ibid.
144 Ibid.
145 *Suo Moto Statement by Shri Yashwant Sinha in Parliament on Insurgents in Bhutan*. For details see <meaindia.nic.in/cgi->.
146 Ibid.
147 'Bhutan Deploys 600 Militias to take on Indian Militants', *The Shillong Times*, 9 December 2003.
148 Ibid.
149 'Bhutan Launched Offensive Against Ultras', *North East Reporter*, 16 December 2003.
150 Ibid.
151 *The Kuensel*, 16 January 2004.
152 Krishna Das, 'Bhutan's Operation All Clear: A Royal Challenge', *North East Sun*, 14 January 2004.
153 *The Kuensel*, 5 June 2004.
154 *The Kuensel*, 14 September 2004.

155 Barun Das Gupta, ‘Dead ULFA Leader Produced’, *The Hindu Online*, 27 December 2003, at <www.hindu.com/2003/12/27/stories/2003122703821200.htm>.
156 *The Kuensel*, 17 December 2004.
157 ‘Bhutan Reassures India of not Allowing ULFA to Camp on its Soil’, dated 25 November 2006, <www.zeenews.com>.
158 Ibid.
159 *The Kuensel*, June 2001.
160 <www.bhutannewsonline.com/economy.html> dated July 4, 2004.
161 *The Kuensel*, 30 March 2005.
162 *The Kuensel*, 29 April 2006, p. 1.
163 Ibid., p. 12.
164 <Bhutannewsonline.htm> and *The Kuensel*, 10 May 2006.
165 *Indian Express*, Chandigarh News line, 13 June 2006.
166 *Asian Development Outlook 2006*, p. 157.
167 Ibid.
168 *The Kuensel*, 14 October 1988.

Bibliography

Ahmad, Zahiruddin, 'The Ancient Frontier of Ladakh', *The World Today*, 16 July 1960.

Aitchison, C. U., *A Collection of Treaties, Engagements and Sanads Relating to India and Neighbouring Countries*, 5th edition, 14 vols. Calcutta: Government of India, Central Publication Branch 1929–33.

Aitchison, J. E. T., *Handbook of the Trade Products of Leh, with the Statistics of the Trade from 1867 to 1872 Inclusive*. Calcutta: Wyman and Co. 1874.

Alder, G. J., *British India's Northern Frontier 1865–95: A Study in Imperial Policy*. London: Longmans 1963.

Alder, Garry, *Beyond Bukhara, the Life of William Moorcroft, Asian Explorer and Pioneer Veterinary Surgeon, 1767–1825*. London: Century Publishers 1985.

Arora, V. K. and A. Appadorai, *India in World Affairs 1957–58*. New Delhi: Oxford University Press 1975.

Bagchi, P. C., *India and China: A Thousand Years of Cultural Relations*. Bombay: Hind Kitab 1950.

Bains, J. S., *India's International Disputes: A Legal Study*. Bombay: Asia Publishing House 1962.

Bajpai, K. Shankar, 'Untangling India and Pakistan', *Foreign Affairs*, New York May–June 2003.

Baral, Lok Raj, *Regional Migration, Ethnicity and Security*. New Delhi: Sterling Publishers 1990.

Bhati, P. S., 'Foreign Policy of Bhutan', *The Indian Journal of Political Studies*, vol. II, December 1987.

Blaikie, Piers *et al.*, *Nepal in Crisis, Growth and Stagnation at the Periphery*. Oxford: Oxford University Press 1980.

Burrad, S. G. and H. H. Hayden, *A Sketch of the Geography and Geology of the Himalaya Mountains and Tibet*. Delhi: Government of India 1933–34.

Chaturvedi, Gyaneshwar, *India–China Relations 1947 to Present*. Agra: M. G. Publishers 1991.

Chuabe, S. K., *The Himalayas: Ethnology and Change*. Berkeley: University of California Press 1985.

Coelho, V. H., *Sikkim and Bhutan*. New Delhi: Indian Council of Cultural Relations 1971.

Constable, Pamela, *My Search for Meaning in the Strife of South Asia*. Washington DC: Brassey's 2004.

Cunningham, Alexander, *Ladak, Physical, Statistical and Historical, with Notices of the Surrounding Countries*. London: W. H. Allen & Co. 1854.

Das, B. S., *Mission to Bhutan: A Nation in Transition*. New Delhi: Vikas Publishing House 1995.

Das, Nirmala, *The Dragon Country*. Bombay: Orient Longman 1974.

Dasgupta, Chandrasekhar, *War and Diplomacy in Kashmir: 1947–1948*. New Delhi: Sage Publishers 2002.

Datta, C. L., *Ladakh and Western Himalayan Politics, 1819–1848*. New Delhi: Munshiram Manoharlal 1973.

Davies, R. H., *Report on the Trade and Resources of the Countries on the North-Western Boundary of British India*. Lahore 1862.

Dilks, David, *Curzon in India*. London: Rupert Hart Davis Ltd. 1970.

Dixit, J. N., *My South Block Years: Memoirs of a Foreign Secretary*. New Delhi: UPSB 1996.

Dixit, J. N., *India's Foreign Policy 1947–2003*. New Delhi: Picus Books 2003.

Drew, Frederic, *The Jummoo and Kashmir Territories*. London: Edward Stanford 1875.

Durrand, Algernon, *The Making of a Frontier*. London: John Murray 1900.

Fisher, James F. (ed.), *Himalayan Anthropology*. The Hague: Mouton 1978.

Fisher, Margaret W. and Leo E. Rose, 'Ladakh and the Sino-Indian Border Crisis', *Asian Survey*, October 1962.

Fisher, Margaret W., Leo E. Rose and R. A. Huttenback, *Himalayan Battleground*. London: Pall Mall 1963.

Francke, A. H., *Antiquities of Indian Tibet*, vol. II. Calcutta: 1926, reprinted Delhi: Chand 1972.

French, Patrick, *Younghusband, The Last Great Imperial Adventurer*. London: Harper Collins Flamingo 1995.

Gaige, Frederick H., *Regionalism and National Unity in Nepal*. New Delhi: Vikas Publishing House 1975.

Gannon, Kathy, 'Afghanistan Unbound'. *Foreign Affairs*, New York, May–June 2004.

Goetz, Hermann, *Studies in the History and Art of Kashmir and the Indian Himalaya*. Wiesbaden: Harrassowitz 1969.

Goldstein, Melvyn C., *A History of Modern Tibet, 1913–1951*. Berkeley: University of California Press 1989.

Grover, B. L. and S. Grover, *A New Look at Modern Indian History*. Delhi: Chand and Company 1995.

Gupta, Sisir, *Kashmir – A Study in India–Pakistan Relations*. Bombay: Asia Publishing House 1966.

Hagen, Toni, *Nepal, the Kingdom in the Himalay*. Berne: Kummerly and Frey 1961.

Hedin, Sven, *Trans-Himalaya, Discoveries and Adventures in Tibet*, 3 vols. London: MacMillan 1910.

Hopkirk, Peter, *The Great Game: The Struggle for Empire in Central Asia*. New York: Kodansha International 1994.

Hopkirk, Peter, *Tresspassers on the Roof of the World: The Secret Exploration of Tibet*. Tokyo and New York: Kodansha International 1995.

Husain, Asad, *British India's Relations with the Kingdom of Nepal*. London: George Allen and Unwin 1970.

Huttenback, Robert A., 'Gulab Singh and the Creation of the Dogra State of Jammu, Kashmir and Ladakh'. *Journal of Asian Studies*, xx(4), August 1961.

Huttenback, Robert A., *British Relations with Sind 1799–1843: An Anatomy of Imperialism*. Los Angeles: University of California Press 1962.

India, Ministry of External Affairs, *Report of United Nations Commission for India and Pakistan*. New Delhi, June 1948 to December 1948.

India, Ministry of External Affairs, *Notes, Memoranda and Letters Exchanged between the Governments of India and China, September to November 1959. White Paper II*. New Delhi 1959.

Jackson, W. A. D., *The Russo-Chinese Borderlands*. Princeton: van Nostrand 1961.

Jha, Prem Shankar, *Kashmir, 1947: Rival Versions of History*. New Delhi: Oxford University Press 1996.

Joshi, Manoj, *The Lost Rebellion: Kashmir in the Nineties*. New Delhi: Penguin Books 1991.

Karan, P. P. and W. Jenkins *Nepal: A Physical and Cultural Geography*. Lexington: University of Kentucky Press 1960.

Karim, Afsar, *Kashmir: The Troubled Frontiers*. New Delhi: Lancer Publishers 1994.

Keay, John, *When Men and Mountains Meet: The Explorers of the Western Himalayas 1820–75*. London: John Murray 1977.

Keay, John, *The Gilgit Game: The Explorers of the Western Himalayas 1865–95*. London: John Murray 1979.

Khalfin, N. A., *Russia's Policy in Central Asia, 1857–1868*. Translated from Russian into English by Hubert Evans. London: 1864.

Khatana, R. P., *Tribal Migration in Himalayan Frontiers: Study of Gujjar Bakarwal Transhumance Economy*. Gurgaon: Vintage Books 1992.

Kohli, Manorama, *India and Bhutan: A Study in Inter-Relations 1772–1910*. New Delhi: Munshiram Manoharlal 1982.

Korbel, Joseph, *Danger in Kashmir*. Princeton: Princeton University Press 1966.

Labh, Kapileshwar, *India and Bhutan*. New Delhi: Sindu Publications 1974.

Lal, J. S. (ed.), *Himalaya: Aspects of Change*. Delhi: Oxford University Press 1981.

Lall, John, *Aksaichin and Sino-Indian Conflict*. New Delhi: Allied Publishers 1989

Lamb, A., *Kashmir: A Disputed Legacy, 1846–1990*. Hertfordshire: Roxford Books 1991.

Lamb, Alastair, *Britain and Chinese Central Asia: The Road to Lhasa 1767–1905*. New York: Routledge and Kegan Paul 1960.

Lamb, Alastair, *The McMohan Line: A Study in the Relations between India, China and Tibet, 1904 to 1914*, vol I: *Morley, Minto and Non-Interference in Tibet*; Vol. II: *Hardinge, McMohan and the Simla Conference*. New York: Routledge and Kegan Paul 1966.

Lamb, Alastair, *The Sino-Indian border in Ladakh*. Canberra: Australian National University Press 1973.

Lattimore, Owen, *Studies in Frontier History, Collected Papers, 1928–58*. New York: Oxford University Press 1962.

Lawrence, Walter, *The Valley of Kashmir*. London: Frowde 1895.

Malla, K. P. (ed.), *Nepal: Perspectives on Continuity and Change*. Kathmandu: CNAS 1989.

Maxwell, Neville, *India's China War*. Bombay: Pelican 1971.

Maynes, Charles William, 'America Discovers Central Asia', *Foreign Affairs*, New York, March/April 2003.

Mehra, P. L., *The Younghusband Expedition: An Interpretation*. Bombay: Asia Publishing House 1969.

Mehra, Parshotam, *An 'Agreed' Frontier*. New Delhi: Oxford University Press 1992.

Mohan, C. Raja, *Crossing the Rubicon: The Shaping of India's New Foreign Policy*. New Delhi: Penguin Books India 2003.

Mukerji, Apratim, *Afghanistan: From Terror to Freedom*. New Delhi: Sterling Publishers: 2003.

Mullik, B. N., *Chinese Betrayal: My Years with Nehru*. Bombay: Allied Publishers 1971.

Muni, S. D., 'Sino-Bhutanese Boundary Talks', *The Indian Express*, 12 June 1984.

Muni, S. D., *India and Nepal: A Changing Relationship*. New Delhi: Konark Publishing Ltd 1992.

National Archives of India (New Delhi), Official Records Proceedings of the Foreign and Political Department, Government of India. *Political Consultations; Political Proceedings; Secret Consultations*.

Pannikar, K. M., *Himalayas in Indian Life*. Bombay: 1963.

Parmanand, *The Politics of Bhutan: Retrospect and Prospect?* New Delhi: Pragati 1992.

Petech, Luciano, 'A Study on the Chronicles of Ladakh, Indian Tibet', *Indian Historical Quarterly*, xv(4), December 1939.

Petech, Luciano, 'The Tibetan-Ladakhi–Mogul War of 1681–83', *Indian Historical Quarterly*, xxiii(3), September 1947.

Petech, Luciano, 'Notes on Ladakhi History', *Indian Historical Quarterly*, xxiv(3), September 1948.

Petech, Luciano, *The Kingdom of Ladakh, c950–1842*. Rome: Instituto Italian per il Medio ed Estremo Oriente 1977.

Prasannan, R., 'Historic Blunders (Suicidal Game)', *The Week*, 30 January 2005.

Raha, M. K. (ed.), *The Himalayan Heritage*. New Delhi: Gian Publishing House 1987.

Rahul, Ram, *The Government and Politics of Tibet*. New Delhi: Vikas Publishing House 1969.

Rajbahak, Ram Prasad, *Nepal–India Open Border: A Bond of Shared Aspirations*. New Delhi: Lancer 1992.

Ramakant and B. C. Upreti (eds), *India and Nepal: Aspect of Interdependent Relations*. Delhi: Kalinga Publications 2001.

Rao, G. Narayan, *The India–China Border – a Reappraisal*. Bombay: Asia Publishing House 1967.

Rathore, L. S., *The Changing Bhutan*. New Delhi: Jain Brothers 1974.

Regmi, D. R., *Ancient Nepal*. Calcutta: Firma K. L. Mukhopadhyay 1960.

Richardson, H. E., *A Short History of Tibet*. New York: EP Dutton and Co. 1962.

Richardson, H. E., *Tibet and its History*. Oxford: Oxford University Press 1962.

Rizvi, Janet, *Ladakh: Crossroads of High Asia*. Delhi: Oxford University Press 1983.

Rose, Leo E., 'Bhutan's External Relations', *Pacific Affairs* 47(2), summer 1974.

Rose, Leo E., *The Politics of Bhutan*. Ithaca and London: Cornell University Press 1974.

Rustomji, Nari, *Bhutan: The Dragon Kingdom in Crisis*. Delhi: Oxford University Press 1978.

Sadhu, J. N., 'Tibetan Refugees in HP Acquiring Property', *Indian Express*, 15 September 1983.

Sanghvi, Ramesh, *India's Northern Frontier and China*. Bombay: Contemporary Publishers 1962.

Scobell, Andrew, *China's Use of Military Force: Beyond the Great Wall and the Long March*. Cambridge: Cambridge University Press 2003.

Sen, L. P., *Slender Was the Thread*. New Delhi: Orient Longman 1994.

Singh, Nagendra, *Bhutan: A Kingdom in the Himalayas*. New Delhi: S. Chand 1972.

Skrine, C. P. and Pamela Nightingale, *Macartney at Kashgar*. Hong Kong: Oxford University Press 1987.

Snellgrove, David, *Buddhist Himalaya*. Oxford: Cassiere 1957.

Snellgrove, David, *The Cultural Heritage of Ladakh*. Delhi: Aris & Phillips 1977.

Stobdan, P., 'Destabilising Presence of Tibetans', *Indian Express*, 8 June 1994.

Stobdan, P., 'Panchen Lama: China's Tibet Policy', *Hindustan Times*, 31 December 1995.

Subba, T. B. and K. Datta (eds.), *Religion and Society in the Himalayas*. Delhi: Gyan Publishing House 1991.

Swinson, Arthur, *North West Frontier People and Events 1839–1947*. London: Hutchinson and Company 1967.

Talbott, Strobe, *Engaging India: Diplomacy, Democracy and the Bomb*. Washington DC: Brookings Institution Press 2004.

Tucci, G., *Ancient Civilisation of Trans-Himalaya*. London: Barrie and Jenkins 1973.

Tyagi, Sushila, *Indo-Nepalese Relations, 1858–1914*. Delhi: D. K. Publishing House 1974.

Upreti, B. C., *Marginal Migrants*. Delhi: Kalinga Publications 2002.

Varma, Yugeshwar P., *Nepal: Progress and Problems 1972–80*. Lalitpur, Nepal: Pratibha Verma 1981.

Vertzberger, Yaacov, *The Enduring Entente: Sino–Pakistan Relations, 1960–80*. New York: Praeger 1983.

Vohra, Sahdev, *The Northern Frontier of India: The Broder Dispute with China*. New Delhi: Intellectual Publishing House 1993.

Warikoo, K., *Jammu, Kashmir and Ladakh: A Comprehensive Bibliography*. New Delhi: Sterling Publishers 1976.

Warikoo, K., *Central Asia and Kashmir: A Study in the Context of Anglo-Russian Rivalry*. Delhi: Gyan Publishing House 1989.

Warikoo, K., 'Ladakh's Trade Relations with Tibet under the Dogras', *China Report*, xxvi(2), 1990.

Warikoo, K. (ed.) *Afghanistan Factor in Central and South Asian Politics*. New Delhi: Trans Asia Informatics 1994.

Warikoo, K. (ed.), *Central Asia: Emerging New Order*. New Delhi: Har Anand 1995.

Warikoo, K. (ed.), *Society and Culture in the Himalayas*. New Delhi: Har Anand 1995.

Warikoo, K. (ed.), *Bamiyan: Challenge to World Heritage*. New Delhi: Bhavna Books 2002.

Warikoo, K. (ed.), *The Afghanistan Crisis: Issues and Perspectives*. New Delhi: Bhavna Books 2002.

Warikoo, K. (ed.), *Afghanistan: The Challenge*. New Delhi: Pentagon Press 2007.

Warikoo, K. and Dawa Norbu (eds.), *Ethnicity and Politics in Central Asia*. New Delhi: South Asian Publishers 1992.

White, J. C., *Sikkim and Bhutan: Twenty-One Years on the North-East Frontier – 1887–1908*. London: Edward Arnold 1909.

Woodman, Dorothy, *Himalayan Frontiers: A Political View of the British, Chinese, Indian and Russian Rivalries*. New York: Praeger 1996.

Index

Yakub Beg 6, 7

www.ingramcontent.com/pod-product-compliance
Lightning Source LLC
LaVergne TN
LVHW021038011026
849160LV00008BA/540
* 9 7 8 0 4 1 5 5 3 3 4 9 2 *